Guide to Human Genome Computing
Second Edition

Guide to Human Genome Computing

Second Edition

Edited by

Martin J. Bishop
Human Genome Mapping Project Resource Centre,
Hinxton, Cambridge, UK

Academic Press
Harcourt Brace & Company, Publishers
San Diego ■ London ■ Boston ■ New York
Sydney ■ Toronto ■ Tokyo

Academic Press
525 B Street, Suite 1900, San Diego, California 92101-4495, USA
http://www.apnet.com

Academic Press Limited
24 – 28 Oval Road, London NW1 7DX, UK
http://www.hbuk.co.uk/ap/

ISBN 0-12-102051-7

First edition published 1994

A catalogue record for this book is available from the British Library

Typeset by Mathematical Composition Setters Ltd, Salisbury, Wiltshire
Printed in Great Britain by The University Press, Cambridge

98 99 00 01 02 03 CU 9 8 7 6 5 4 3 2 1

Contents

Contributors   ix
Preface   xiii

1 **Introduction to Human Genome Computing Via the
   World Wide Web**   1
   Lincoln D. Stein

   1   Introduction   1
   2   Equipment for the tour   2
   3   Genome databases   3
   4   Analytic tools   24
   5   Conclusion   38
   6   URLs   39

2 **Biological Materials and Services**   41
   Michael Rhodes and Ramnath Elaswarapu

   1   Introduction   41
   2   Genomic resources available   42
   3   Chromosome-specific libraries   46
   4   CpG island libraries   46
   5   cDNA libraries   47
   6   Probe and primer banks   47
   7   Cell lines   48
   8   Cell hybrid panels   48
   9   Backcross panels   48

3 **Managing Pedigree and Genotype Data**   49
   Stephen P. Bryant

   1   Introduction   49
   2   Fundamentals   50
   3   Modelling the domain   55
   4   Implementing the database   59

5 Interfacing with analysis programs   64
6 Displaying and managing data   73
7 Further development   73

4 Linkage Analysis Using Affected Sib-Pairs   75
Pak Sham and Jinghua Zhao

1 Introduction   75
2 Genetic identity-by-descent between sib-pairs   75
3 Genetic identity-by-descent and recombination fraction   77
4 Identity-by-descent distribution for an affected sib-pair   78
5 Methods of affected sib-pair analysis   81
6 Affected sib-pair analysis: strengths and limitations   84
7 Conclusion   85
List of computer programs   87

5 Comparative Mapping in Humans and Vertebrates   89
Martin Bishop

1 Conservation of genes   89
2 Conservation of linkage groups   92
3 Value of comparative mapping   94
4 Single-species databases   95
5 Protein families   107
6 Comparative mapping databases   109

6 Radiation Hybrid Mapping   113
Linda McCarthy and Carol Soderlund

1 Introduction   113
2 Background   114
3 RH mapping concepts   115
4 Radiation hybrid mapping programs   119
5 Conclusions   145

7 Sequence Ready Clone Maps   151
Carol Soderlund, Simon Gregory and Ian Dunham

1 Introduction   151

2	Strategy	153
3	Software for sequence ready maps	155
4	Long-range map construction (SAM, Z-RHMAPPER)	156
5	STS-content data storage using ACEDB	159
6	Sequence ready contig construction using FPC	160
7	Process tracking and coordination	172
8	Conclusions	177

8 Software for Human Genome Sequencing   181
Simon Dear

1	Introduction	181
2	Overview	182
3	Gel image processing	182
4	Sequence preprocessing	185
5	Sequence assembly	189
6	Editing and finishing	192
7	Feature identification	197
8	Quality control	200
9	The finished sequence	200
10	Conclusion	201
11	Finding software on the Web	202

9 Human EST Sequences   205
Guy St C. Slater

1	Introduction	205
2	Background	205
3	Generation of the EST data	206
4	Reasons for EST sequencing	207
5	Publicly available EST data	209
6	EST analysis tools	211
7	Biological resources	212
8	WWW resources for ESTs	213
9	Summary	213

10 Prediction of Human Gene Structure   215
Luciano Milanesi and Igor B. Rogozin

| 1 | Introduction | 215 |
| 2 | Genome structure | 216 |

3   Approaches to evaluate prediction accuracy   217
4   Functional sites in nucleotide sequences   219
5   Functional regions in nucleotide sequences   230
6   Protein coding gene structure prediction   236
7   Analysis of potential proteins coded by predicted genes   251
8   RNA-coding gene structure prediction   252
9   Sequence analysis   252

11 Gene Finding: Putting the Parts Together   **261**
    Anders Krogh

1   Introduction   261
2   Dynamic programming   262
3   State models   266
4   Performance comparison   271
5   Conclusion   272

12 Gene-Expression Databases   **275**
    Duncan Davidson, Martin Ringwald
    and Christophe Dubreuil

1   Introduction   275
2   The scope of gene-expression databases   277
3   A survey of gene-expression data   278
4   A survey of gene-expression databases   284
5   The future of gene-expression databases   292

Index   **295**

Contributors

Martin J. Bishop
Medical Research Council, UK Human Genome Mapping Project Resource Centre, Hinxton, Cambridge, CB10 1SB, UK
Tel: 01223 494500, Fax: 01223 494512

Stephen P. Bryant
Gemini Research Limited, 162 Science Park, Milton Road, Cambridge, CB4 4GH, UK
E-mail: Steve.Bryant@gemini-research.co.uk
Tel: 01223 435307, Fax: 01223 435301

Duncan Davidson
Developmental Genetics Section, MRC Human Genetics Unit, Western General Hospital, Crewe Road, Edinburgh EH4 2XU, UK
Tel: 0131 332 2471, Fax: 0131 343 2620

Simon Dear
The Sanger Centre, Wellcome Trust Genome Campus, Hinxton, Cambridge, CB10 1SA, UK
E-mail: sd@sanger.ac.uk
Tel: 01223 834244, Fax: 01223 494919

Christophe Dubreuil
Developmental Genetics Section, MRC Human Genetics Unit, Western General Hospital, Crewe Road, Edinburgh EH4 2XU, UK
Tel: 0131 332 2471, Fax: 0131 343 2620

Ian Dunham
The Sanger Centre, Wellcome Trust Genome Campus, Hinxton, Cambridge, CB10 1SA, UK
Tel: 01223 494948, Fax: 01223 494919

Ramnath Elaswarapu
UK Human Genome Mapping Project Resource Centre, Hinxton, Cambridge, CB10 1SB, UK
Tel: 01223 494540, Fax: 01223 494512

Simon Gregory
The Sanger Centre, Wellcome Trust Genome Campus, Hinxton, Cambridge, CB10 1SA, UK
Tel: 01223 494948, Fax: 01223 494919

Anders Krogh
Center for Biological Sequence Analysis, Technical University of Denmark, Building 206, 2800 Lyngby, Denmark
Tel: +45 4593 4808, Fax: +45 4525 2470

Linda McCarthy
Department of Genetics, University of Cambridge, Downing Site, Tennis Court Road, Cambridge, CB2 3EH, UK
Tel: 01223 333999, Fax: 01223 333992

Luciano Milanesi
Istituto Tecnologie Biomediche Avanzate, CNR, Via Fratelli Cervi 93, 20090 Segrate, Milano, Italy
E-Mail: milanesi@itba.mi.cnr.it
Tel: +39 26422706, Fax: +39 26422770

Michael Rhodes
UK Human Genome Mapping Project Resource Centre, Hinxton, Cambridge, CB10 1SB, UK
Tel: 01223 494540, Fax: 01223 494512

Martin Ringwald
The Jackson Laboratory, 600 Main Street, Bar Harbor, ME 04609, USA

Igor B. Rogozin
Institute of Cytology and Genetics SD RAS, Lavrentyeva av. 10, Novosibirsk 630090, Russia and Istituto Tecnologie Biomediche Avanzate, CNR, via Fratelli Cervi 93, 20090 Segrate, Milano, Italy

Pak Sham
Department of Psychological Medicine, Institute of Psychiatry, De Crespigny Park, Denmark Hill, London SE5 8AF, UK
Tel: 0171 919 3888, Fax: 0171 701 9044

Guy St C. Slater
UK Human Genome Mapping Project Resource Centre, Hinxton, Cambridge, CB10 1SB
Tel: 01223 494500, Fax: 01223 494512

Carol Soderlund
The Sanger Centre, Wellcome Trust Genome Campus, Hinxton, Cambridge, CB10 1SA, UK
Tel: 01223 494948, Fax: 01223 494919

Lincoln D. Stein
Whitehead Institute, MIT Center for Genome Research, 9 Cambridge Center, Cambridge, MA 02142-1497, USA
E-mail: lstein@genome.wi.mit.edu
Tel: +1 617 252 1900, Fax: +1 617 252 1902

Jinghua Zhao
Department of Psychological Medicine, Institute of Psychiatry, De Crespigny Park, Denmark Hill, London SE5 8AF, UK
Tel: 0171 919 3536, Fax: 0171 701 9044

Preface

The first stage of the Human Genome Project, sequencing of representative DNA to the extent that the only gaps are repetitive sequences or regions which present major technical problems in cloning or sequencing, should be completed by the middle of the next decade. The methodology to do this is in place and is described in this book in the context of bioinformatics. The work is in progress at a number of centres worldwide. The expected major technical advances necessary to speed the process have so far not been forthcoming. It is business as usual in factory scale operations.

Genome projects involve a systematic approach to the identification of all the genes of an organism. Comparative approaches involve yeast, nematode, fly, fish (fugu and zebra), frog, chicken and mouse (and others). Access to the physical DNA via overlapping clone maps has proved remarkably difficult to provide in man. Radiation hybrid mapping and BAC libraries have eased the situation.

A high resolution genetic linkage map has been constructed for the mouse. CEPH families and sib-pairs are the major tools for genetic linkage studies in man but the resolution is lower than for mouse. The mouse is therefore crucial in relating phenotypes to genes.

cDNA (EST) sequencing and mapping on radiation hybrid DNA promise to raise the number of sequenced, mapped and named human genes from 6000 to 50 000 in a short time. A further 50 000 genes encoding transcripts which are expressed in low amounts or are highly localized in time (development) or space (specialized cell types) are expected to be discovered by genomic sequencing. Expression profiles of cells can be studied by massively parallel methodologies. Pathways involving metabolic and developmental genes are being elucidated and need to be modeled. Understanding the function of 10 000 proteins will take longer but commercial rewards are available today.

Bioinformatics is the key to the organization and retrieval of data on genome organization and macromolecular function. Databases, query and browsing tools and analysis tools are evolving world wide to meet the challenge. Many of these are highly graphical. The spectacular rise of the World Wide Web to be the major access method has occurred since the First Edition was published and we provide a wide ranging and clear introduction (Chapter 1). Many tasks are still too time-consuming and labour intensive. The aim for future development is to reduce human intervention in analysis and reporting without loss of validity of results. Reliability of predictions will be enhanced as knowledge of sequence, structure and function advances.

Our subject has very major implications for mankind in terms of agriculture, biotechnology, pharmaceuticals and general healthcare. The ethical issues which are emerging will require very sensitive and intelligent treatment.

MJ Bishop

1 Introduction to Human Genome Computing Via the World Wide Web

Lincoln D. Stein

Whitehead Institute, MIT Center for Genome Research, 9 Cambridge Center, Cambridge, MA 02142-1497, USA. E-mail: lstein@genome.wi.mit.edu

1 INTRODUCTION

Since the first edition of this book was published, the genome community has seen an explosion in the number and variety of resources available over the Internet. In large part this explosion is due to the invention of the World Wide Web, a system of document linking and integration that has gone from obscurity to commonplace in a mere five years.

The genome community was an early adopter of the Web, finding in it a way to publish its vast accumulation of data, and to express the rich interconnectedness of biologic information. The Web is the home of primary data, of genome maps, of expression data, of DNA and protein sequences, of X-ray crystallographic structures, and of the genome project's huge outpouring of publications. These data, spread out among thousands of individual laboratories and community databases, are hot-linked throughout. Researchers who wish to learn more about a particular gene can (with a bit of patience) move from physical map to clone, to sequence, to disease linkage, to literature references and back again, all without leaving the comfort of their Web browser application.

However, the Web is much more than a static repository of information. The Web is increasingly being used as a front end for sophisticated analytic software. Sequence similarity search engines, protein structural motif finders, exon identifiers, and even mapping programs have all been integrated into the Web. Java applets are adding rapidly to Web browsers' capabilities, enabling pages to be far more interactive than the original *click–fetch–click* interface. It may soon be possible for biologists to do all their computational work with no more than a browser on their desktop computers.

Guide to Human Genome Computing, 2nd edition   Copyright © 1998 Academic Press Limited
ISBN 0-12-102051-7

This chapter is an illustrated tour of the World Wide Web from the genome biologist's perspective. It does not pretend to be a technical discussion of Web protocols or to explain how things work. Nor is there any attempt for this to be an exhaustive listing of all the myriad Web resources available. Instead I have attempted to show the range of resources available and to give guidance on how to learn more. Other chapters in this book delve more deeply into selected topics of the Web and genome.

URLs (Universal Resource Locators) are necessary for interactive Web browsing but terribly ugly when they appear on the printed page. I have gathered all the URLs and placed them in a table at the end of this chapter. Within the body of the text I refer to them by descriptive names such as 'Pedro's Home Page' rather than by their less friendly addresses.

2   EQUIPMENT FOR THE TOUR

2.1   Web Browser

The Web was designed to run with any browser software, on any combination of hardware and operating system. However, the pace of change has outstripped many software developers. Although some Web sites can still be viewed with older browsers (such as the venerable National Center for Supercomputing Applications (NCSA) Mosaic or the Windows Cello browser), many require advanced features found only in recent browsers from the Microsoft and Netscape companies. For most effective genome browsing, I recommend one of the following browsers:

1. Netscape Navigator, 3.02 or higher
2. Netscape Communicator 4.01 or higher
3. Microsoft Internet Explorer, 3.01 or higher

These browsers can be downloaded free of charge from the Netscape and Microsoft home pages. They are also available from most computer stores and mail order outfits.

Although it is good to have a recent version of the browser software installed, it may not be such a good idea to use the *most* recent, as these versions often contain bugs that cause frustrating crashes. Be particularly wary of 'pre-release', 'preview', and 'beta' browser versions.

2.2   Internet Connection

A direct connection to the Internet is a necessity for Web browsing. All academic centers, government laboratories and nearly all private companies usually have a fast Internet connection of at least 56K bps (56 000 bits per second). This will be more than adequate for Web browsing purposes.

Home users usually dial into the Internet via an Internet Service Provider using a PPP (point-to-point protocol) or SLIP (serial line interface protocol) connection. For such users a modem of 28.8K bps or better is strongly recommended.

3   GENOME DATABASES

3.1   The WWW Virtual Library: Genetics

Although we could start our tour anywhere, a good place to begin is the genetics division of the *WWW Virtual Library*, a distributed topic-oriented collection of Web resources (Figure 1.1). This page contains links to several hundred sites around the world, organized by organism.

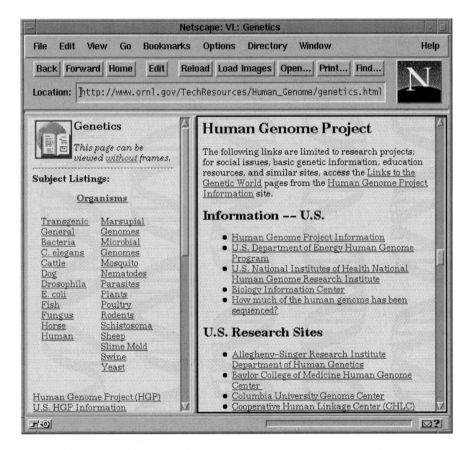

Figure 1.1   *The WWW Virtual Library, a good jumping-off point for genome resources on the Web.*

The list of organisms in the left-hand frame provides a quick way to jump to the relevant section. Click on the link labeled 'Human' to see sites under this heading. Subheadings direct you to a variety of US and international sites, as well as to chromosome-specific Web pages and search services.

3.2   Entrez

We select the link for *GenBank*, taking us to the home page of the National Center for Biotechnology Information (Figure 1.2). NCBI administers GenBank, the main repository for all published nucleotide sequencing information. Links from its home page will take you to GenBank, as well as to SwissProt, the protein

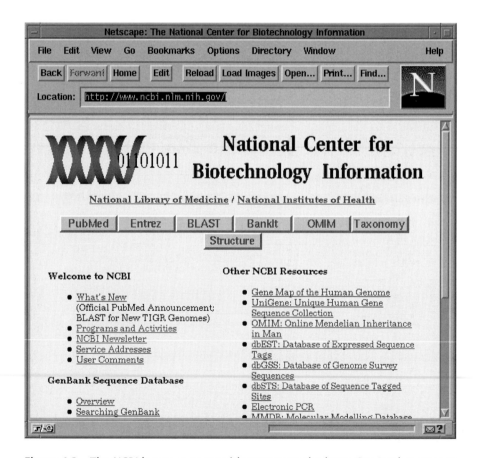

Figure 1.2   *The NCBI home page provides access to the huge GenBank sequence database.*

sequence repository, OMIM, the Online Mendelian Inheritance in Man collection of genetic disorders, MMDB, a database of crystallographic structures, and several other important resources.

While there are several ways to access GenBank and the other databases, the most useful interface is the *Entrez* search engine, an integrated Web front end to many of the databases that NCBI supports. To access Entrez, we click on the labeled button in the navigation bar at the top of the window.

This takes us to the Entrez welcome page shown in Figure 1.3. The links on this page point to Entrez's six main divisions:

1. *PubMed Division.* This is an interface to the MedLine bibliographic citation service. Some nine million citations of papers in the biologic and biomedical literature are available going back as far as 1966. Most citations are accompanied by full abstracts.

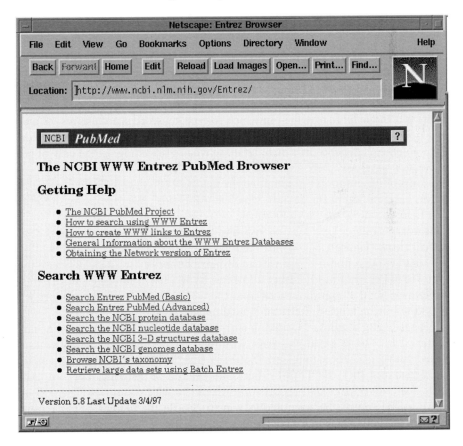

Figure 1.3   *The Entrez search engine provides access to GenBank's bibliographic, nucleotide, protein, structural and genome divisions.*

2. *Nucleotide Database.* This is the GenBank collection of nucleotide sequences, now merged with the EMBL database.
3. *Protein Database.* This database combines primary protein sequencing data from SwissProt and other protein database collections, together with protein sequences derived from translated GenBank entries.
4. *3D Structures Database.* This contains protein three-dimensional (3D) structural information derived from X-ray crystallography and nuclear magnetic resonance (NMR). The source of the structures is the MMDB (Molecular Modelling Database) maintained at Brookhaven National Laboratories.
5. *Genomes Database.* This is a compilation of genetic and physical maps from a variety of species. Maps of similar regions are integrated to allow for comparisons among them.
6. *Taxonomy.* This is the phylogenetic taxonomy used throughout GenBank. Its primary purpose is as a consultation guide to obscure species.

A search on the nucleotide division will illustrate how Entrez works. For this example, we'll say that we're interested in information on the 'sushi' family of repeats found in many serine proteases. Selecting the link labeled 'Search the NCBI nucleotide database' displays a page similar to the one shown in Figure 1.4. There is a single large text field in which to type keyword search terms, as well as a pop-up menu that allows us to limit the search to certain database fields. Available fields depend on which database we are searching. In the case of the nucleotide database, there are nearly two dozen fields covering everything from the name of the author who submitted the sequence entry to the sequence length. In this case we accept the default, which is 'All Fields'. We type 'sushi' into the text field and press the 'Search' button.

Searches rarely take longer than a few seconds to complete. The page that appears now (Figure 1.5) indicates that eight entries matched our search. This is a small enough number that we can display the entire list. In cases where too many matches are found, Entrez allows us to add new search terms, progressively narrowing down the search until the number of hits is manageable. We press the button labeled 'Retrieve 8 documents'.

A page listing a series of GenBank entries that match the search now appears (Figure 1.6). This is a complex page with multiple options. Each entry is associated with a checkbox to its left. You may select all or a subset of the entries on the list and click the 'Display' button at the top of the page. This will generate a summary page that reports on each of the selected entries. The pop-up menu at the top of the page allows you to choose the format of the report. Choices include the standard GenBank format, a list of bibliographic references for the selected entries, the list of protein 'links', and the list of nucleotide 'neighbors' (more on 'links' and 'neighbors' later).

You may also retrieve information about a single entry. Following the

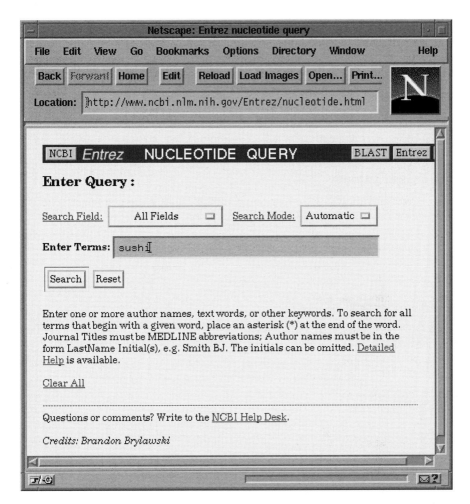

Figure 1.4 *Searching the nucleotide database for entries that refer to 'sushi'.*

description there are a series of hypertext links, each linking to a page that gives more information about the entry. Depending on the entry, certain links may or may not be present. A brief description of these links is as follows:

1. *GenBank report.* This shows the raw GenBank entry in the form that most biologists are familiar with.
2. *Sequence report.* This is the GenBank entry in a friendlier text format.
3. *FASTA report.* This is just the nucleotide sequence in the format accepted by the FASTA similarity searching program.
4. *ASN.1 format.* A structured format used by the NCBI databases (and almost no one else).

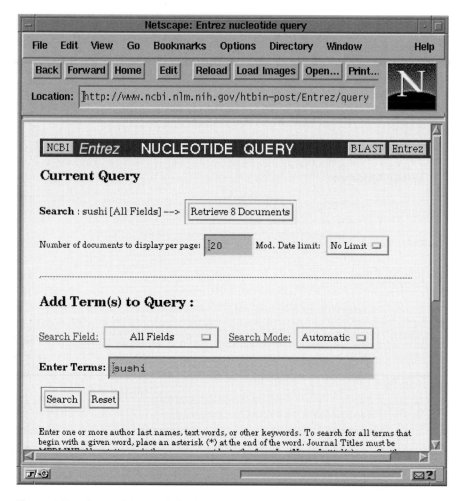

Figure 1.5 *The 'sushi' search finds eight documents. We can either view them or refine the search further.*

5. *Graphical view*. For sequences derived from cosmids, bacterial artificial chromosomes (BACs) and other contigs, this shows a graphic representation of the sequencing strategy.

6. *NN genome links*. If the entry corresponds to a sequence that has been placed on one or more physical or genetic maps, this link appears. Selecting it will jump to the Entrez Genomes division (see p.11). The number of maps that the entry appears in will replace the 'NN'.

7. *NN Medline links*. If a published paper refers to the entry, this link appears. Selecting it will jump to the list of paper(s) in the Entrez bibliographic division.

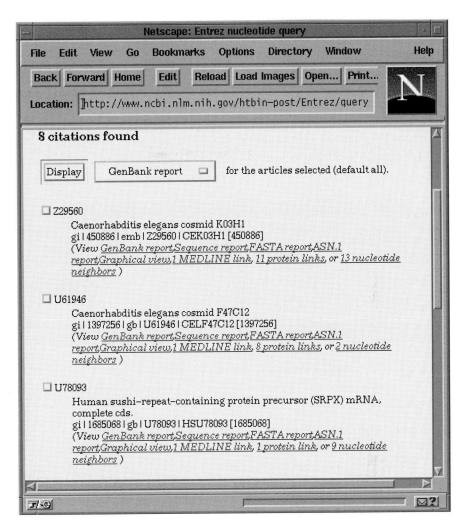

Figure 1.6   *Entrez presents search results as a list of hotlinks to GenBank entries.*

8. *NN structural links.* As above, but for 3D structures (see later).
9. *NN protein links.* This corresponds to protein sequences related to the entry. If the entry contains an open reading frame (real or predicted) there will be at least one protein link.
10. *NN nucleotide neighbors.* Each nucleotide entry added to GenBank is routinely BLASTed against all previous entries (see later) to create precalculated sets of 'neighbors' that share sequence similarity. If a nucleotide entry has any sequence-similarity neighbors, this link will appear.

To continue our example, we decide to investigate GenBank entry U78093, described as a Human 'sushi-repeat-containing protein precursor'. Selecting '1 MEDLINE links' takes us to a page that lists the one paper that refers to this entry (not shown), and prompts us to select its citation format. Selecting the default format displays the citation shown in Figure 1.7. This article indicates that the gene in question is deleted in some patients with retinitis pigmentosa, and offers us links (in the form of buttons) to related articles, other relevant DNA and protein sequences, and to entries in OMIM that deal with retinitis pigmentosa.

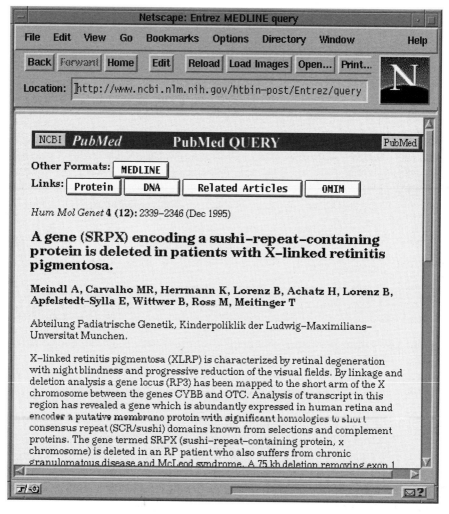

Figure 1.7   *The GenBank entry for accession number U78093.*

Returning to our original list of sushi sequences, we can now select the link labeled '9 nucleotide neighbors'. This takes us to a list of all the sequence entries in GenBank that have significant BLAST homologies to U78093. Here we find several EST (expressed sequence tags) entries produced by the Washington University/Merck cDNA sequencing project. It is possible that some of them represent previously undescribed members of the sushi family of serine proteases.

The user interface for other Entrez divisions provides a similar search–link–follow interface. The exception is the genomes division, which, because it has fewer entries than the others, is entered through a straightforward listing of prominent organisms and the genome maps available for them. From the Entrez welcome page, we select 'Search the NCBI genomes database' and then 'Homo sapiens' from the list of prominent organisms (not shown). This leads us to a list of 26 maps (22 autosomes, one sex chromosome and three mitochondrial maps) from which we select human chromosome 14. This leads us to a page (Figure 1.8), that displays a single prominent image in the center.

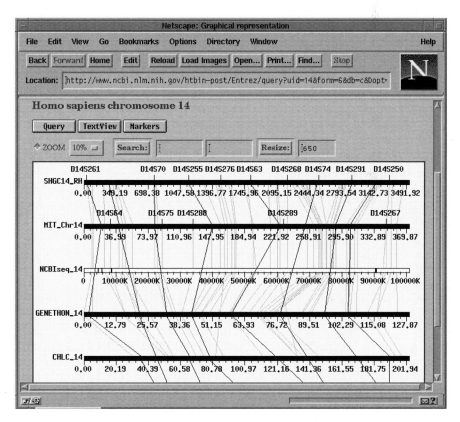

Figure 1.8 *The genomes division of Entrez has a graphic interface based on alignments among multiple maps.*

The image shows a series of genetic and physical maps published from a variety of sources, roughly aligned, with diagonal lines connecting common features. The image is 'live', meaning that we can click on it to magnify areas or to view information about individual maps. When the magnification is large enough to see individually mapped objects (sequences, genetic loci and sequence tagged sites (STSs)), clicking on them will take us to a page showing the object's GenBank record, where we can learn more about it in the manner described above.

If you are interested in a known physical or genetic region and wish to view it directly, the genomes division interface allows you to type in the names of the two mapped loci that define the region. The map will be expanded and scrolled to the proper area. You can then examine the map for interesting candidate genes near the region of interest.

Entrez's 3D structures division contains entries for several thousand proteins and other macromolecules whose structures have been determined by X-ray crystallography and/or NMR. The entries are fully linked to related entries in the nucleotide, protein, citation and genome divisions.

To get the most out of the 3D structures, you will need to install a 'helper application' to view and explore the MMDB structure files. Entrez supports two different helpers, Rasmol and Kinemage. Both are available in versions that run on Macintosh, Windows and Unix systems. You will need to obtain and install one of these software packages, then configure your browser to launch it automatically to view a structure file. Full instructions can be found at Entrez's MMDB FAQ (frequently-asked questions) page.

The search interface to the 3D structures division is nearly identical to that used for nucleotide and protein sequences. You enter one or more keywords into a text field and press the 'Search' button, optionally limiting the scope of the search to a particular field. However, the retrieved entries will contain two links that we have not seen before, *Structure Summary* and *NN structure neighbors*. The first link retrieves a page that describes the entry's structure in a standardized format. The second link indicates the presence of one or more entries that are structurally 'similar' to the entry. 'Similarity', in the case of 3D structures, is determined by an algorithm that measures the volume of overlap between the two molecules.

Searching for the term 'sushi' in this case was ineffective, but searching for 'serine protease' was more productive, recovering 136 entries with structural information. Selecting the 'Structure Summary' link for any of the matching entries retrieves a page that gives information on the structural determination method and its citation. A series of pop-up menus and push buttons allows you to retrieve the 3D structure in a variety of formats. Selecting 'RasMol' format (assuming that the RasMol viewer is installed) and pressing the 'View' button launches the helper application (Figure 1.9). You are now free to rotate the image with the mouse, magnify it, adjust various display options, and save the structure to local disk for further exploration.

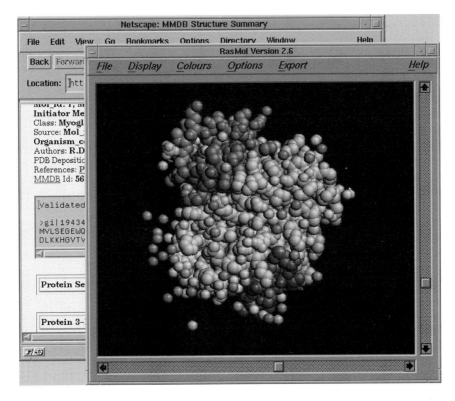

Figure 1.9 *Entrez's structural division uses external viewers to display and rotate 3D protein models.*

3.3 'Gene Map' and UniGene Databases

No tour of the NCBI's Web site is complete without a side trip to the 'Gene Map of the Human Genome', a compendium of approximately 16 000 expressed sequences from the UniGene set that have been localized by radiation hybrid mapping (see Chapter 6). These maps were published in late 1996 by a consortium of research groups. Although the maps are already somewhat out of date, it is expected that these pages will be updated at regular intervals.

From the NCBI home page, select the link labeled 'Gene Map of the Human Genome'. This leads you to a pastel page that offers a series of ideograms of human chromosomes. There are several ways to search this database. If the region you are interested in is defined cytogenetically, just click on the ideogram in the desired region. A page like that shown in Figure 1.10 will appear showing a list of all mapped expressed sequences in the area. Selecting the GenBank

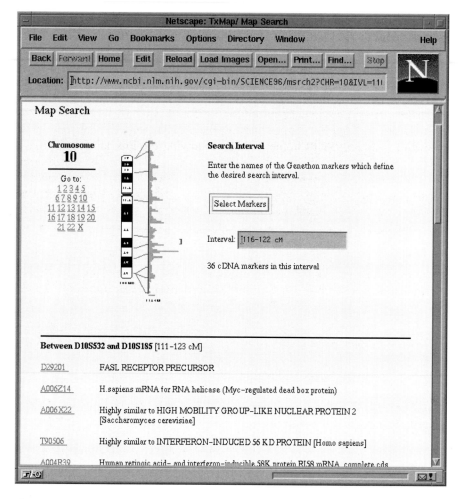

Figure 1.10   *The NCBI gene map allows you to search for expressed genes by name or position.*

accession numbers of the retrieved sequences will bring up pages with further information about the sequences and how they were mapped. If the region of interest is defined by markers on the Genethon genetic map, you can search for all expressed sequences located between any pair of markers.

There is also a more flexible, but less obvious, interface to the gene map database. Look for an inconspicuous link labeled 'Research Tools Page' at the bottom of the gene map home page. This will lead you to a page that links to various types of searches, including text-based searches and sequence searches. The latter search is of particular interest. It prompts you to paste a new unknown sequence into a text field. When you press the search button, the NCBI server performs a BLAST sequence similarity search against all the expressed sequences on the gene map. This is a rapid way to find sequence similarities to previously mapped expressed sequences, and may be helpful in certain positional cloning strategies.

Closely related to the gene map is the NCBI *UniGene* set, a collection of human transcripts that have been clustered in an attempt to create a set of unique expressed sequences (see Chapter 9). The UniGene set was compiled from two primary sources, random cDNA sequencing efforts from Washington University and elsewhere (dbEST), and published genes from GenBank. UniGene can be browsed from the NCBI home page by following the link labeled 'UniGene: Unique Human Gene Sequence Collection'. There is both a map-oriented search facility, which allows you to list expressed sequences that have been placed on a particular chromosome, and a keyword search facility. The difference between the map search facilities offered in the Gene Map and UniGene pages is that the latter includes expressed sequences that have been assigned to chromosomes but not otherwise ordered.

Both search interfaces will eventually generate a listing of matching UniGene clusters, along with a short phrase that describes each one. Select the clusters of interest to see the individual dbEST and GenBank entries that comprise the set. You can then browse individual sequence entries.

3.4 SRS

Although Entrez might appear to be the be-all and end-all source of protein and nucleotide sequence information, this is not quite true. There are many smaller but well curated databases of biologic information that are not included among the databases that Entrez serves. These include the Prosite and Blocks databases of protein structural motifs, transcription factor databases, species-specific databases, and databases devoted to certain pathogens.

The SRS (Sequence Retrieval System) is a Web-based system for searching among multiple sequence databases supported by the European Molecular Biology Laboratory (EMBL). In addition to the large EMBL sequence database, it cross-references sequence information from approximately 40 other sequence databases (the precise number is slowly increasing). Among these databases are ones that hold protein and nucleotide sequence information, 3D structure, disease and phenotype information, and functional information.

The SRS system is replicated among multiple sites in order to distribute network load. You can access it directly from its home page in Heidelberg, Germany, or follow a link from this page to locate the site nearest you (there are servers in Europe, Asia, the Pacific and South America; North America is conspicuously absent). To begin a search, connect to an SRS site (Figure 1.11). You can use the Heidelberg server or one of its replicated sites, several of which you will find via links in the WWW Virtual Library. Closer sites may have better response times.

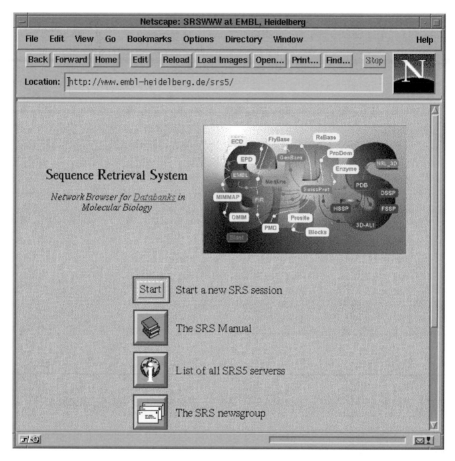

Figure 1.11   *The SRS sequence search system links 40 different molecular biology databases.*

The SRS home page contains links that you can follow to learn more about the SRS service. Click 'Start' to begin a new SRS searching session.

You will be asked to select the databases to search (not shown). There are 40-odd checkboxes on this page, each corresponding to a different source database. This may seem formidable at first, but fortunately the databases have been grouped by category into nucleotide sequence related, protein related, and so on. Check off the databases that you are interested in searching. If you are not familiar with a particular database, click its name to obtain a brief description. In our example, we will search the motif databases for proteins containing the zinc finger motif. We select the Prosite and Blocks databases and press 'Continue' to move to the next page.

This takes us to a page similar to that shown in Figure 1.12. The most

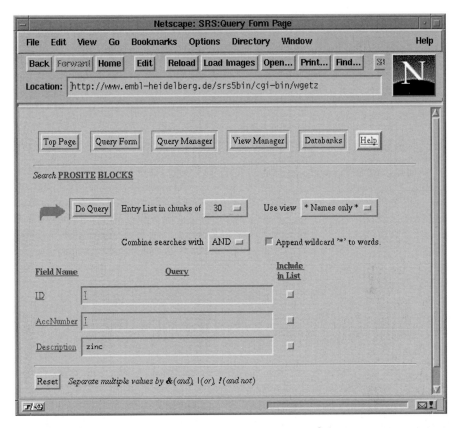

Figure 1.12 *SRS search pages allow you to perform structured (field-based) queries on one or more databases.*

important elements are a set of text fields at the bottom of the page, each one corresponding to a different field in the database(s) selected. The number and nature of the fields will depend on which databases are selected. If multiple databases are selected, SRS displays only the fields that are shared in common among them all. This can be a trap for the unwary: if you select too many databases to search at once, you may find that the only field displayed is the (usually unhelpful) entry ID field. Page back and deselect some databases, or uncheck the option labeled 'Show only fields that selected databanks have in common'.

Our example shows three fields labeled 'ID', 'AccNumber' and 'Description'. We can click on the name of each field to learn more about it, but in this case it seems obvious that 'Description' is the one we want. We type in 'zinc' and press the 'Do Query' button.

This search results in a list of 65 matches (Figure 1.13) to entries in one or

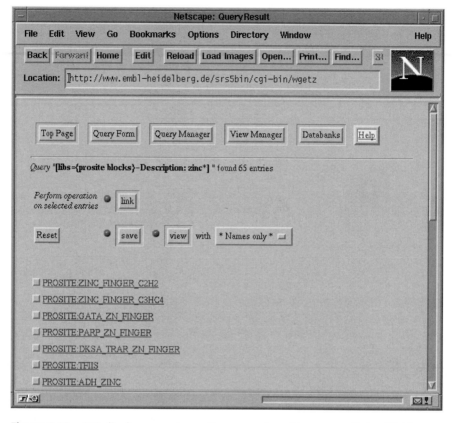

Figure 1.13   *SRS displays search results as a series of hypertext links. Clicking on the buttons at top broadens the search to other databases by bringing in cross-references.*

more of the selected databases. The format of each match is the name of the database, e.g. *PROSITE*, followed by a colon and the ID of the entry. We can now:

1. Select an entry by clicking on its name, fetching information about it.
2. Expand the search by selecting the checkboxes to the left of one or more entries and then choosing one of the buttons labeled 'link', 'save' or 'view'.

If we click on an entry's name, we will be taken to its database record. What exactly is displayed will depend on the structure of the database. For members of the Prosite database, the record consists of a description of the structural motif and a list of all the entries in GenBank/EMBL that are known to contain this motif.

The more interesting operation is to select the checkboxes of a series of entries, then press the 'link' button. This performs a search of the other databases in the SRS system and returns all entries that are cross-referenced with the selected entries. In our example, we select the C2H2 and C3HC4 zinc finger domains and press 'link', taking us to a page that prompts us to select the databases to link to (not shown). We select the 'SwissProt', 'EMBL' and 'GenBank' sequence databases and press 'continue'.

The resulting page (Figure 1.14) lists 861 matches. We can scan through individual entries, or repeat the linking process to expand the scope of the search still further.

Other options on the search results page allow you to create and download reports on the selected matches in a variety of formats. Select the preferred format

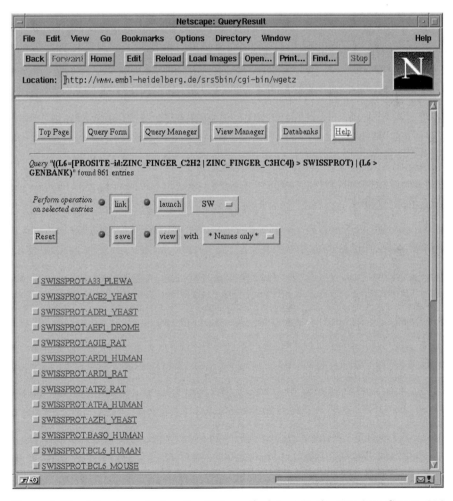

Figure 1.14   *After broadening the SRS search shown in the previous figure, SRS now brings in entries from SwissProt and other databases.*

from the page's pop-up menu, and click either the 'save' button to download the report to your local disk, or 'view' to see the report into the browser. The report options available to you depend on the databases you have selected. Some databases offer a simple text-only report, others offer more options. For example, the protein databases offer a fancy on-screen Java hydrophobicity chart. Other reports offer the ability to search the databases by sequence similarity using either the FASTA or Smith–Waterman algorithms.

3.5   GDB

GDB, the Genome Database, is the main repository for all published mapping information generated by the Human Genome Project. It is a species-specific database: only *Homo sapiens* maps are represented. Among the information stored in GDB is:

- genetic maps
- physical maps (clone, STS and fluorescent *in situ* hybridization (FISH) based)
- cytogenetic maps
- physical mapping reagents (clones, STSs)
- polymorphism information
- citations

To access GDB, connect to its home page (Figure 1.15). GDB offers several different ways to search the maps:

- *A simple search*. This search, accessible from GDB's home page, allows you to perform an unstructured search of the database by keyword or the ID of the record. For example, a keyword search for 'insulin' retrieves a list of clones and STSs that have something to do either with the insulin gene or with diabetes mellitus.
- *Structured searches*. A variety of structured searches available via the link labeled 'Other Search Options' allow you to search the database in a more deliberate manner. You may search for maps containing a particular region of interest (defined cytogenetically, by chromosome, or by proximity to a known marker) or for individual map markers based on a particular attribute (e.g. map position and marker type). GDB also offers a 'Find a gene' interface that searches through the various aliases to find the gene that you are searching for.

Searches that recover individual map markers and clones will display them in a list of hypertext links similar to those displayed by Entrez and PDB. When

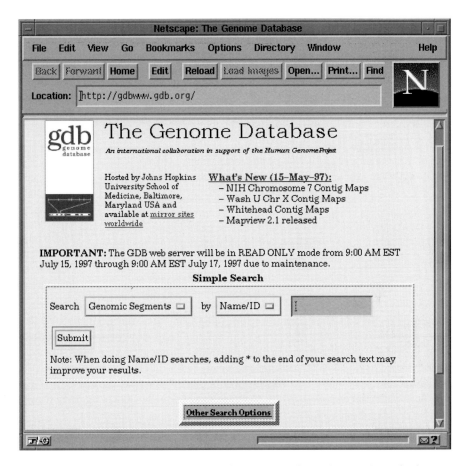

Figure 1.15   *The GDB home page provides access to the main repository for human genome mapping information.*

you select an entry you will be shown a page similar to Figure 1.16. Links on the page lead to citation information, information on maps this reagent has been assigned to, and cross-references to the GenBank sequence for the marker or clone. GDB holds no primary sequence information, but the Web's ability to interconnect databases makes this almost unnoticeable.

A more interesting interface appears when a search recovers a map. In this case, GDB launches a Java applet to display it. If multiple maps are retrieved by the search, the maps are aligned and displayed side by side (Figure 1.17). A variety of settings allows you to adjust the appearance of the map, as well as to turn certain maps on and off. Double clicking on any map element will display its GDB entry in a separate window.

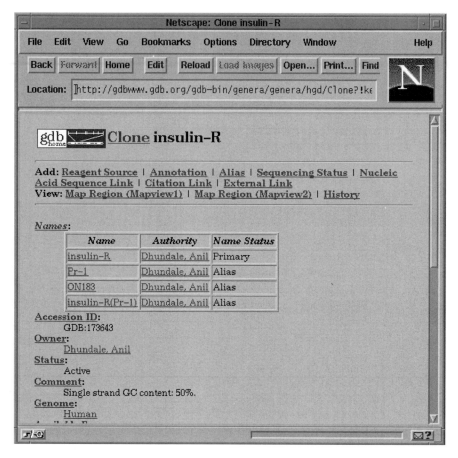

Figure 1.16   *GDB displays most entries using a text format like that shown here.*

3.6   Species-Specific Databases

In addition to the large community databases like GenBank, EMBL and GDB, there are hundreds of smaller species-specific databases available on the Web. Although not offering the comprehensive range of the big databases, they are a good source of unfiltered primary data. In addition they may be more timely than the community databases because of the inevitable lag between data production and publication.

Notable sites in this category include:

■   *Whitehead Institute/MIT Center for Genome Research.* The data available at this Web site include genome-wide genetic and physical maps

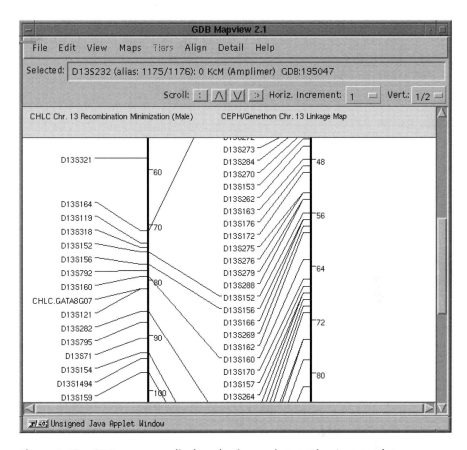

Figure 1.17   *GDB maps are displayed using an interactive Java applet.*

of the mouse, physical maps of the human, a genetic map of the rat, and human chromosome 17 DNA sequence.

■ *MGD (Mouse Genome Database)*. This database, based at the Jackson Laboratory, contains mouse physical and genetic mapping information, DNA sequencing data, and a rich collection of mouse strains and mutants.

■ *Stanford Human Genome Center*. This is the site of an ongoing project to produce a high-resolution radiation hybrid map of the human genome.

■ *FlyBase*. This Web database, hosted at the University of Indiana, is a repository of maps, reagents, strains and citations for *Drosophila melanogaster*.

■ *ACEDB*. The ACEDB database stores mapping, sequencing, citation and developmental information on *Caenorhabditis elegans* and other organisms. The Genome Informatics Group at the University of

Maryland maintains a Web site at the URL given below, which provides interfaces both to the *C. elegans* database and to a variety of plant, fungal and prokaryotic genomes.

■ *SGD (Saccharomyces Genome Database)*. The Stanford Genome Center hosts the Saccharomyces Genome Database, a repository of everything that is worth knowing about yeast (now including the complete DNA sequence).

■ *TIGR (The Institute for Genome Research)*. The TIGR site contains partial and complete genomic sequences of a large number of prokaryotic, fungal and protozoan organisms. Its 'Human Gene Index' is a search interface to the large number of human expressed sequences that have been produced by TIGR and other groups.

■ *Washington University Genome Sequencing Center*. This is the home of several large-scale genome sequencing projects, including human and mouse EST sequencing, *C. elegans* genomic sequencing, and human genomics sequencing (primary chromosomes 2 and 7).

■ *The Sanger Centre*. The Sanger Centre is another source of extensive DNA sequencing information. Projects include the genomic sequence of *C. elegans*, and human chromosomes 1, 6, 20, 22 and X. In addition to its sequencing efforts, the Sanger Centre also produces chromosome-specific human radiation hybrid maps.

■ *University of Washington*. The University of Washington Genome Center is sequencing human chromosome 7 (in collaboration with Washington University), as well as the human leukocyte antigen (HLA) class I region and the mouse T-cell receptor region.

4   ANALYTIC TOOLS

We now turn our attention to the analytic tools available on the Web. A few years ago, the simplest type of sequence analysis was hindered by the need to find the right software for your computer, install it, learn the ins and outs of the interface, and format the data according to the program's needs. This has become much simpler recently as more and more of the standard computational tools in the molecular biologist's armamentarium have been put on-line with easy to use Web front ends. This section tours the tools that are currently available for the most frequent types of analysis.

4.1   BLAST Searches

The most basic computational tool is the BLAST search, a rapid comparison of a search sequence to a database of known sequences. This search is used routinely to determine whether a newly sequenced DNA has already been published in the literature, and, if not, to give some hint of its putative function by searching for related sequences.

Many Web sites offer BLAST search interfaces, including SRS (discussed above) and the NCSA Biologist's Workbench (discussed below). Probably the most widely used interface is that offered by the NCBI. To use this interface, connect to the NCBI BLAST search page and select the type of search to perform. NCBI offers both 'basic' and 'advanced' searches. The first uses sensible default parameters for the search. The latter allows you to fine-tune the BLAST search parameters, something that is recommended only if you fully understand what you are doing (extensive on-line documentation on the BLAST algorithm is available at the NCBI site). Because of the recent release of large amounts of TIGR-specific EST data at the time of writing, NCBI was also offering BLAST searches restricted to the TIGR dataset. This may no longer be available by the time you read this.

For the purposes of our tour, we will select the link pointing to the 'basic' search, arriving at the page shown in Figure 1.18. The BLAST interface allows us to search for sequences using one of several different algorithms selected from a pop-up menu near the top of the page. The algorithms are described in more detail in the on-line documentation, but I summarize them here for convenience:

- *blastn.* The search sequence is compared directly to the database sequences, using parameters appropriate for nucleotides.
- *blastp.* The search sequence is compared directly to the database sequences, using parameters appropriate for protein sequence.
- *blastx.* The search nucleotide sequence is first translated into protein sequence in all six reading frames, then compared against a database of protein sequences.
- *tblastn.* The search protein sequence is compared against a database of nucleotide sequences after translating each database entry into protein using all six reading frames.
- *tblastx.* The search nucleotide sequence is compared against a database of nucleotide sequences, after translating both the search sequence and the database sequences into protein in all six reading frames.

The advantage of using one of the *blastx*, *tblastn* or *tblastx* search methods is that you are able to find matches to distantly related sequences. The disadvantage is that the searches become computationally intensive and may take an inordinate length of time.

In addition to selecting the algorithm, you will also be asked to select the database to search. You may search the default 'NR' database, which contains a list of all non-redundant nucleotide sequences known to GenBank, or restrict the search to various species-specific collections, ESTs, or to new entries submitted during the past month. We leave the default at 'NR'.

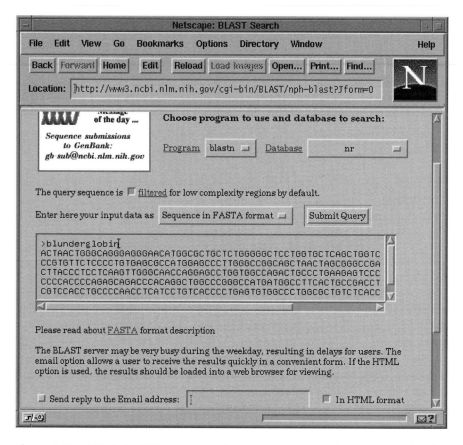

Figure 1.18 *NCBI's BLAST page provides rapid sequence similarity searches for both protein and nucleotide sequences.*

The next task is to enter the sequence itself. The BLAST interface offers two ways to specify a sequence. You may cut and paste the sequence directly into the large text field in the center of the page. Alternatively, if the search sequence is already a part of GenBank, you can select 'Accession or GI' from the pop-up menu above the text field and enter the sequence's GenBank accession number in the text area.

If you choose to enter the raw sequence, you must be careful to use FASTA format. This format begins with the name of the sequence on the top line, preceded by a '>' sign. Following this is the sequence itself, which should contain no line numbers or spaces. The figure shows an example of a valid search sequence.

We fill in the text field with our search sequence ('blunderglobin') and press the 'Submit Query' button located above the text area. A few seconds later our query returns with a list of possible matches (Figure 1.19) ordered so that the

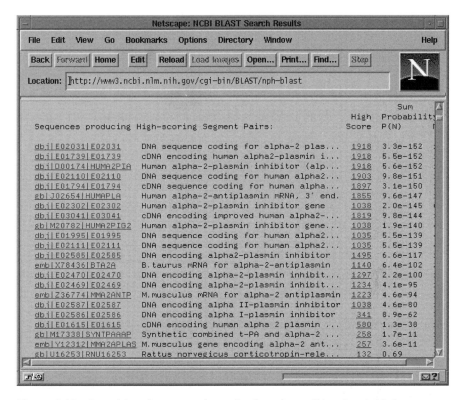

Figure 1.19   *Searching for a match to the imaginary 'blunderglobin' sequence using BLAST.*

most similar sequences are located at the top. We can now click on the links for the matches to view their GenBank entries.

The BLAST server may be slow during periods of heavy usage. At the bottom of the search page a pair of checkboxes and an additional 'Submit Query' button allow you to have the BLAST server send the results of the search to your e-mail address. This allows you to launch several searches without waiting for each one to complete.

4.2   Primer Picking

Another bread-and-butter task is to pick a PCR primer pair from a DNA sequence in order to create an STS (sequence tagged site). The Whitehead Institute/MIT Center for Genome Research provides a handy on-line primer picking tool, an interface to its freeware PRIMER program. To use this resource, connect to the Whitehead's home page and select the link labeled 'WWW Primer Picking'. The interface is straightforward (Figure 1.20). Paste the DNA sequence into the large text field at the top of the form and press the button labeled 'Pick Primers'. The sequence should contain only the characters AGCTN and white space. Case is ignored. Although the program can handle large sequences, it is wise not to paste in sequences much longer than you need. There is no need to enter 20K of sequence in order to generate an STS 200 bp long.

In this example, we have pasted in the sequence for our unknown 'blunderglobin' gene. After pressing the primer-picking button, the program offers us five sets of primer pairs that define STSs ranging in size from 117 to 256 bp. We can accept these or page back to the previous page to change the primer picking parameters. The default parameters pick primers that satisfy PCR conditions used at the Whitehead and many other laboratories. However, all parameters are adjustable. You may adjust the PCR conditions, the preferred PCR product size, and the stringency of the primer picking. You may also designate regions that the program shall exclude from primer picking, or which it will

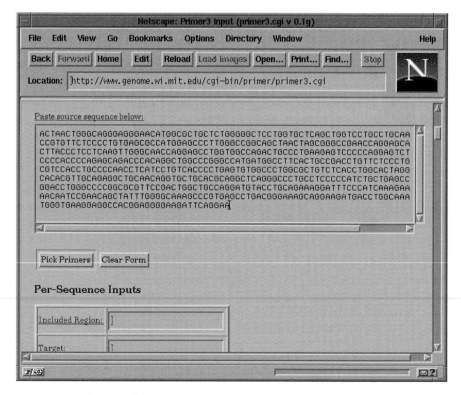

Figure 1.20   *Primer picking with the Whitehead Institute's PRIMER tool.*

attempt to include in the PCR product. Another option allows you to pick a third oligonucleotide within the PCR product for the purpose of certain protocols that use hybridization to detect the product.

4.3   Exon Prediction

As the rate of genomic DNA sequencing has increased, it has become ever more important to have tools that can predict the presence of genes in genomic sequence. While far from perfect, these exon prediction tools (also known as 'gene finders' and 'sequence annotators') can give you a first start at finding the location of potential genes.

The Baylor College of Medicine's *Gene Finder* program is the most straightforward of the exon prediction programs. To use it, connect to the Baylor Molecular Biology Computational Resources page and follow the links labeled 'Services on the Web' and 'The BCM Genefinder' (not shown). Paste the sequence into the large text field at the top of the page, and enter its name in the small text field where indicated. For certain long-running algorithms you are also asked to enter your e-mail address so that the results can be sent to you off-line. Sequences of up to 7 kb are accepted. For longer sequences, the Gene Finder page instructs you to use an e-mail interface instead.

This is the easy part. The hard part is choosing the exon prediction algorithm and parameter set from among the 20-odd possibilities that Gene Finder's page offers. While specific instructions are given in the search page, the default of 'FGENEH' is most suitable for human genomic sequences. Other algorithms are better tuned for invertebrates, prokaryotes and fungi.

To test Gene Finder, we paste the first 4K of the human apolipoprotein CI sequence into the field and press 'Perform Search'. The result is shown below:

```
Name: ApoCI
First three lines of sequence:
TATCGCATGCAGCCCCCAGTCACGCATCCCCTGCTTGTTCAATCGATCACGACCCTCTCACGTGCACCCACTTAG
AGTTGTGAGCCCCTTAAAAGGAACAGGGATTGCTCACTCGGGGAGCTCGGCTCTTGAGACAGGAATCTTGCCCATT
CCCCGAACGAATAAACCCCTTCCTTCGTTAACTCAGCGTCTGAGGAATTTTGTCTGCGGCTCCTCCTGCTACATT
fgeneh  Fri Jul 11 11:14:04 CDT 1997   ApoCI
   Nucleotides which are not A,C,G,T,R or Y were removed from your sequence.
   length of sequence -   2401
   number of predicted exons -  3
   positions of predicted exons:
      468 -     529 w=   7.68
      664 -     741 w=  13.76
     1986 -    2296 w=   5.94
   Length of Coding region-  451bp    Amino acid sequence - 149aa
```

```
LIKVLRAGQDLPTKPSSKDSECPSGLAMRLFLSLPVLVVVLSIVLEGPAPAQGTPDVSSA
LDKLKEFGNTLEDKARELISRIKQSELSAKMRLEPFPGHGRAGVCFWVEPWQMVQDEQIE
KKTSPGEADNIPLVTQLDLKVLRLQGQFP*
```

In this case Gene Finder identified three potential exons in this sequence. The second two correspond to known exons in the ApoCI gene. The first, however, is a false hit. It spans an area overlapping an untranscribed area and the 5′ untranslated region. A 60% accuracy rate is typical of the current generation of exon identification tools.

In addition to the exon prediction service, Baylor offers a number of on-line tools for molecular biologists, including protein secondary structure prediction, sequence alignment, and a service that launches sequence similarity searches on a number of databases.

Another exon predictor is GRAIL (Gene Recognition and Assembly Internet Link), a service provided by the Oak Ridge National Laboratory. The GRAIL engine can detect other features in addition to exons, including poly-adenylation sites, repeat sequences and CpG islands. The interface is relatively simple (Figure 1.21). Choose the feature you wish to search for, and paste the sequence into the text field at the bottom of the page (scrolled out of view in the screenshot). Alternatively, a 'file upload' button allows you to load the sequence directly from a text-only file on your local disk. The exact nature of each feature that GRAIL can search for is described in detail in the program's on-line manual.

Selecting 'Grail 2 Exons' from the list of features and repeating our experiment with ApoCI gave the following results:

```
[grail2exons - Exons]

      St Fr Start      End ORFstart ORFend      Score     Quality
  1-  f  2    664      741     636     821    100.000    excellent
  2-  f  1   1986     2121    1814    2296     98.000    excellent

[grail2exons - Exon Translations]

3- SPEPLPLPPECPSGLAMRLFLSLPVLVVVLSIVLEGKSGMGELGS

4- FEPLPIFLAGPAPAQGTPDVSSALDKLKEFGNTLEDKARELISRIKQSEL
   SAKMRLEPFPGHGR
```

In this case, the two correct exons were identified.

4.4   NCSA Biology Workbench

The NCSA was responsible for Mosaic, the graphical Web browser that set off

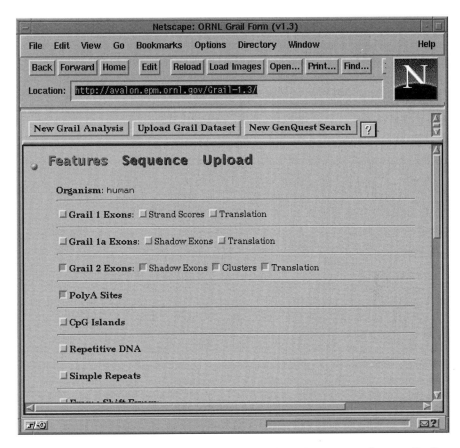

Figure 1.21   *The GRAIL site provides on-line nucleotide sequence feature-finding services. The checkboxes allow you to select which features to search for.*

the explosion of interest in the Web. It is also responsible for the *Biology Workbench*, an integrated package of several dozen protein and nucleotide sequence search and analysis tools.

To use Workbench you will need to register an account with the Workbench server. This is because Workbench allows you to save personal project data on the server itself. Your account name and password provide a way to return to the data and ensure a degree of privacy. To create an account, go to the Biology Workbench home page (not shown) and select the link labeled 'Account Set-Up'. You will be prompted for a log-in name and password.

After the account is created, you will be able to enter the service by following the link labeled 'Welcome to the NCSA Biology Workbench'. Figure 1.22 shows the main Biology Workbench screen. The menu bar at the top of the page contains five subdivisions labeled 'Session Tools', 'Protein Tools', 'Nucleic Tools',

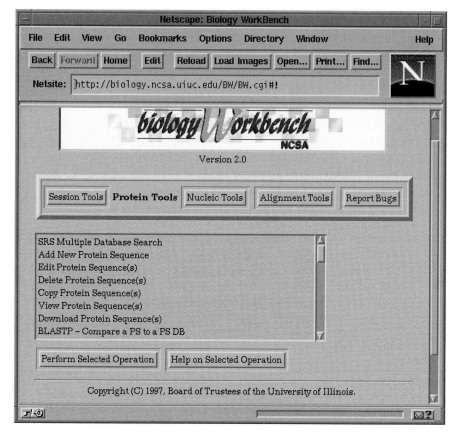

Figure 1.22   *The NCSA Biology Workbench has four main analytic subdivisions, selected by using the menu buttons at the top.*

'Alignment Tools' and 'Report Bugs'. The Protein, Nucleic acid and Alignment tool buttons lead to pages that run various analytical programs. 'Session tools' allows you to save your work to a named 'session', log your actions, and restore an old session at some later date. The meaning of 'Report Bugs' should be obvious.

Workbench is confusing at first because of its many options. Once you understand its style, however, it is easy to use. The general strategy for a Workbench session is as follows:

1. Import sequences into Workbench.
2. Select one or more sequences to analyze.
3. Select the analysis to perform.
4. Run the analysis.

4.4.1   Importing Sequences

Select the type of analysis you wish to perform from the main Protein, Nucleic Acid and Alignment divisions. A scrolling list of possible analyses will appear (Figure 1.22), among which are options to add a new sequence and perform an SRS database search. The first option allows you to import a new sequence into the workbench by cutting and pasting into your browser. The second is an interface to the SRS search engine (see the section above). Sequences recovered from the SRS search engine can be imported into Workbench without an intermediate cut-and-paste step. The interfaces for these two options are similar to ones we have seen already. Once a sequence has been imported into Workbench, you can view it, edit it, or delete it.

4.4.2   Selecting One or More Sequences to Analyse

Once sequences have been imported into Workbench, they will appear in a list below the menu bar (see Figure 1.23). To the left of each sequence is a checkbox. To select the sequence(s) to analyse, just check the appropriate box.

4.4.3   Selecting the Analysis to Perform

The list of analyses spans the spectrum from sequence similarity searches to alignments to protein secondary structure prediction. What is displayed in the scrolling list depends on which of the major subdivisions you have selected. Only one analysis can be selected at a time. Some analyses require only one sequence only to be selected, while others require two or more. In the example shown in Figure 1.23, we have imported and selected both the mRNA and genomic sequences for the human ApoCI gene and will be using the CLUSTALW algorithm to obtain a sequence alignment.

4.4.4   Run the Analysis

Press the button labeled 'Perform Selected Operation'. Depending on the analysis, you may now be asked to view and adjust some of its parameters.

The format of the output depends on the analysis. In the case of the attempted alignment between the genomic and mRNA ApoCI sequences, Workbench produces a large text file showing the expected alignment between the two sequences. A button at the bottom of the output prompts us to import this alignment file back into Workbench. Doing so gives us an 'alignment' object that we can then view with one of the Workbench's alignment display tools (Figure 1.24).

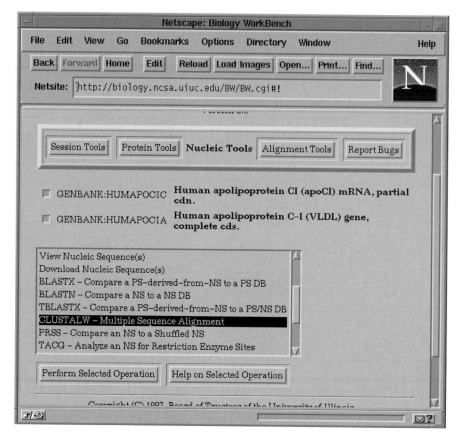

Figure 1.23   *Performing an analysis with Biology Workbench is a matter of selecting the sequences to analyze and the analytic program to run.*

Interestingly (but not too surprisingly), the coding sequence that GenBank's two entries give for this gene are not quite the same. Regrettably there is no existing on-line artificial intelligence service that will help sort out this type of problem.

4.5   Physical Mapping Tools

Several Web sites offer you the ability to map new STSs to one or more of the existing physical maps of the human genome. Two of the more useful services are that of the Whitehead Institute/MIT Center for Genome Research, which offers mapping services for its radiation hybrid map, and that of the Stanford

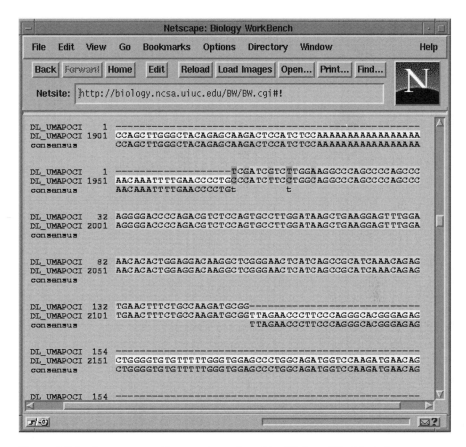

Figure 1.24 *A sequence alignment produced by Biology Workbench.*

Human Genome Center, which will place STSs on its high-resolution radiation hybrid map. Used in conjunction with the primer picking service described above, both services provide you with a rapid way to map new clones.

To use either of the radiation hybrid mapping services, you must obtain DNAs from the same radiation hybrid screening panel that was used to construct the map (see Chapter 6 for full details on radiation hybrid mapping). The Whitehead map was constructed from the Genebridge 4 RH panel, whereas the Stanford map used the higher-resolution G3 mapping panel. DNAs are available from a number of biotech supply houses, including Research Genetics of Huntsville, Alabama. Each STS to be mapped must be amplified on the DNAs from the hybrid panel, then scored on agarose or acrylamide gel. For best results, all amplifications should be done in duplicate; results that are discrepant should be repeated or treated as unknown.

To place STSs on the Whitehead map, reformat the hybrid panel screening results in standard 'radiation hybrid vector' format. The format looks like this:

```
sts_name1
001001011000001000000011010001101110011100101001211001101
sts_name2
000001111000001000000011010000001110011100101001211001101...
```

Each digit is the result of the polymerase chain reaction (PCR) on one of the radiation hybrid cell lines. '0' indicates that the PCR was negative (no reaction product), '1' that it was positive, and '2' is used for 'unknown' or 'not done'. The order of digits in the vector is important, and must correspond to the official order of the Genebridge 4 radiation hybrid panel. The correct order is given in the help page of the Whitehead server (see below), and is identical to the order in which the DNAs are packaged when they are shipped by Research Genetics. You can place spaces within the vector in order to increase readability. The STS name should be separated from the screening data with one or more spaces or tabs.

From the Whitehead home page, follow the link labeled 'Map STSs relative to the human radiation hybrid map' (Figure 1.25). Enter your e-mail address where indicated, and cut and paste the PCR scores into the large text field at the top of the page. It is important that you enter the correct e-mail address, as this is the only way in which you can be informed of the mapping results.

By default, the mapping results are returned in text form. If you wish to generate graphic pictures of the STSs placed on the Whitehead map, you must select the desired graphics format. Currently the PICT and GIF formats are available. The former is appropriate if you are using a Macintosh system. The latter is appropriate for Windows and UNIX systems. Select the graphics format by choosing the appropriate radio button from the labeled set (scrolled out of site in the figure).

When you are satisfied with the settings, press the 'Submit' button. You will receive a confirmation that the data have been submitted for mapping. The results will be returned to you via e-mail shortly (if the server is heavily loaded, however, it may take several hours). If the STS was successfully mapped, the e-mail will list the chromosome it linked to, and its position relative to other markers on the Whitehead map. If requested, you will also receive a picture of the map (with the location of the newly mapped STSs marked in red) as an e-mail enclosure.

The Whitehead also offers access to its STS content-based physical map of the human genome. If you have screened one or more STSs against the CEPH mega-YAC library (see Chapter 2), you can use a search page located at the Whitehead site to determine which YAC contigs contain the YACs hit by your STSs. From this you can infer the position of the STSs relative to the Whitehead

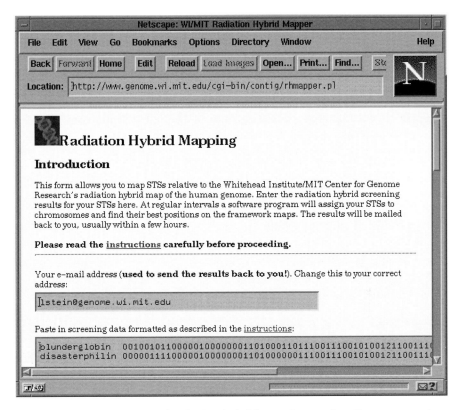

Figure 1.25 *The Whitehead radiation hybrid mapping service allows you to place new STSs on the Whitehead radiation hybrid map by pasting in PCR amplification data.*

map. You can access this service by connecting to the Whitehead home page. Then follow the links labeled 'Human Physical Mapping Project' and 'Search for a YAC by its address'.

To place STSs on the Stanford RH map prepare your data in a similar way, but using the G3 mapping panel. The other important difference is that the PCR result vectors should use an 'R' rather than a '2' to indicate missing or discrepant data. For the Stanford service, data vectors should not contain white space.

Connect to the Stanford Genome Center's home page and follow the links to 'RH Server' and then to 'RHServer Web Submission'. Enter your e-mail address and a reference number in the indicated fields. The e-mail address is vital to receive the mapping results. The reference number is an optional field that will be returned to you with the results and is intended to help you keep the results organized. If known, also enter the STS's chromosomal assignment into the field labeled 'Chromosome number'. This information increases the ability of the mapping software to detect a valid linkage.

Now cut and paste the screening results into the large text field and press the 'Submit' button. Mapping results are typically returned via e-mail within a few minutes. The Stanford server returns the mapping results as a series of placements relative to genetic markers. For each STS, the server reports the closest genetic marker, its chromosome, and the distance (in centiRays) from the marker to the STS. Although no graphic display is provided for the mapping results, the retrieved information can be used in conjunction with the browsable maps available at the Stanford site in order to infer the location of the newly mapped STS relative to other STSs on the Stanford radiation hybrid map.

4.6   Other Analytic Tools

There are many more resources for computational biology than we could possibly fit into this tour. Fortunately the Web makes it easy to find them. They are never more than one or two jumps away from the following pages:

- *Pedro's BioMolecular Research Tools.* This is a vast compendium of computational biology tools. Here you will find both the on-line sort and the traditional sort that must be downloaded and installed. One caveat: this page has not been updated for over a year and may be out of date by the time you read this.
- *Baylor College of Medicine Computational Resources.* The BCM site that we visited earlier contains a large number of links to on-line analytic tools around the world.
- *Whitehead Institute Biocomputing Links.* The Whitehead Institute maintains a well organized and up-to-date listing of analytic tools for molecular biology.
- *Dan Jacobson's Archive of Molecular Biology Software.* This page lists links to nearly 100 repositories of molecular biology software. If you cannot find it here, it probably does not exist.

5   CONCLUSION

The Web has already revolutionized the way that biologists work and, to some extent, think. The next few years will see even more dramatic changes as the databases, analytic tools, and perhaps even the software used to acquire and manage primary data merge together. As large-scale genomic sequencing swings into full gear, there will be a need for tools to allow physically distant collaborators to edit and annotate sequences, run analyses, share their results with the community, and take issue with other laboratories' findings. The Web will provide the essential infrastructure for those tools.

We can look forward to some interesting times ahead.

6   URLs

6.1   Web Browsers

Netscape Corporation   http://www.netscape.com/
Microsoft Corporation   http://www.microsoft.com/

6.2   Genome Databases

WWW Virtual Library, Genetics Division
   http://www.ornl.gov/TechResources/Human_Genome/genetics.html
GenBank/National Center for Biotechnology Information (NCBI)
   http://www.ncbi.nlm.nih.gov/
Entrez Home Page
   http://www3.ncbi.nlm.nih.gov/Entrez/
Entrez MMDB FAQ Page
   http://www3.ncbi.nlm.nih.gov/Entrez/struchelp.html
NCBI Gene Map of the Human Genome
   http://www.ncbi.nlm.nih.gov/SCIENCE96/
SRS Home Page
   http://www.embl-heidelberg.de/srs5/
GDB Home Page
   http://gdbwww.gdb.org/
Whitehead Institute/MIT Center for Genome Research
   http://www.genome.wi.mit.edu/
MGD Home Page
   http://www.jax.org/
Stanford Human Genome Center
   http://shgc.stanford.edu/
FlyBase
   http://flybase.bio.indiana.edu/
ACEDB
   http://probe.nalusda.gov:8300/other/

Saccharomyces Genome Database
   http://genome-www.stanford.edu/Saccharomyces/
The Institute for Genome Research
   http://www.tigr.org/
Washington University Genome Sequencing Center
   http://genome.wustl.edu/gsc/gschmpg.html
The Sanger Centre
   http://www.sanger.ac.uk/
University of Washington
   http://chimera.biotech.washington.edu/uwgc/

6.3   Analytic Tools

NCBI BLAST Searches
   http://www3.ncbi.nlm.nih.gov/BLAST/
Whitehead PCR Primer Picking
   http://www.genome.wi.mit.edu/
Baylor College of Medicine Computational Resources
   http://condor.bcm.tmc.edu/home.html
GRAIL
   http://avalon.epm.ornl.gov/Grail-1.3/
NCSA Biology Workbench
   http://biology.ncsa.uiuc.edu/
Direct Link to Whitehead Radiation Hybrid Mapping Service
   http://www.genome.wi.mit.edu/cgi-bin/contig/rhmapper.pl
Pedro's BioMolecular Research Tools
   http://www.public.iastate.edu/~pedro/research_tools.html
Whitehead Institute Biology Resources
   http://www.wi.mit.edu/bio/biology.html
Dan Jacobson's Archive of Molecular Biology Software
   http://www.gdb.org/Dan/softsearch/biol-links.html

2 Biological Materials and Services

Michael Rhodes and Ramnath Elaswarapu

UK Human Genome Mapping Project Resource Centre, Hinxton, Cambridge CB10 1SB, UK

1 INTRODUCTION

As already detailed in the previous chapter the Internet has become a vital part of any biologist's toolkit. In recent years a wide range of sites offering bio-informatics resources has sprung up. As well as providing rapid access to a huge range of information sources, the Internet and particularly the World Wide Web (WWW) is being used increasingly as a means of obtaining genomic reagents and services. Two main types of supplier are using the WWW: commercial suppliers who use it as an integral part of their marketing strategy, and the research organizations using the WWW to advertize their services. It is the intention of this chapter to provide a, necessarily brief, introduction to some of these services and the means to access them. Some services have become so ubiquitous that they will not be described (e.g. DNA Sequencing, Primer Synthesis). It should be noted that some of the academic suppliers have restrictions on the use of resources supplied. The suppliers listed in this chapter have minimal restrictions, usually concerned with commercial use of the resources. Users should check on any restrictions before obtaining resources.

It is the primary intention of this chapter to give information on the availability of human resources but, owing to their importance, model organisms will also be covered. First the different resources will be described and then two tables will be provided that give the location of various resources. Table 2.1 will list the various resources available and who can supply them. Table 2.2 will provide details of these suppliers.

Guide to Human Genome Computing, 2nd edition   Copyright © 1998 Academic Press Limited
ISBN 0-12-102051-7

Table 2.1a   Human biological resources

Service	Suppliers
Genomic libraries	
YAC libraries	CEPH, Gen Sys, Res Gen, RZPD, HGMP-RC
BAC libraries	Res Gen, Gen Sys, RZPD
PAC	Gen Sys, Res Gen, RP, RZPD, HGMP-RC
Chromosome-specific libraries	
Cosmid libraries	RIKEN, RZPD, HGMP-RC
Phage libraries	ATCC
YAC	Res Gen, RZPD, HGMP-RC
BAC	CEPH, CSM
CpG island libraries	HGMP-RC
cDNA libraries	
Specific libraries	JCRB, Res Gen, RIKEN, RZPD, HGMP-RC
IMAGE libraries	ATCC, Gen Sys, Res Gen, RZPD, HGMP-RC
Probe and primer banks	
Probe banks	ATCC, ECACC, JCRB, HGMP-RC
Primer banks	ATCC, Res Gen, HGMP-RC
Cell lines	ATCC, CCR, ECACC, Genethon, RIKEN
Cell hybrids	
Somatic cell hybrids	Bios, CCR, Res Gen, HGMP-RC
Radiation hybrids	HGMP-RC, Res Gen

This table lists the type of service and major suppliers. See Table 2.2 for the full name and details of suppliers. Note: Suppliers are often not the actual originators of the resources. All suppliers provide links to or information on the resources provided.

2   GENOMIC RESOURCES AVAILABLE

2.1   Genomic Libraries

The purpose of a genomic library is to give the researcher access to the total genome of the organism being studied. Libraries are presented in a variety of vectors and hosts. Libraries are usually distributed to the user as high-density spotted filters (up to 150 000 clones per filter) or polymerase chain reaction

Table 2.1b   Model organism resources available

Service	Suppliers
Genomic libraries	
YAC libraries	Gen Sys (mouse), Res Gen (mouse, rat), RIKEN (mouse), RZPD (mouse, rat, zebrafish, pig), HGMP-RC (mouse)
BAC libraries	Gen Sys (mouse, rat, *Drosophila*, zebrafish), Res Gen (mouse), RZPD (mouse)
PAC	Gen Sys (mouse, rat, *Drosophila*, zebrafish), ResGen (mouse, rat), RP (mouse, rat), RZPD (mouse)
Cosmid	RZBD (fugu), HGMP-RC (fugu)
CpG island libraries	HGMP-RC (mouse)
cDNA libraries	
Specific libraries	RZPD (mouse, zebrafish, *Xenopus*), HGMP-RC (mouse)
IMAGE libraries	ATCC (mouse), Gen Sys (mouse), Res Gen (mouse), RZPD (mouse), HGMP-RC (mouse)
Probe and primer banks	
Primer banks	ATCC (mouse), Res Gen (mouse, zebrafish, rat, *Xenopus*), HGMP-RC (mouse)
Cell lines	JCRB (mouse), RIKEN (mouse)
Cell hybrids	
Somatic cell hybrids	HGMP-RC (mouse)
Radiation hybrids	Res Gen (mouse, baboon, zebrafish)
Backcross mapping panels	Jackson (mouse), HGMP-RC (mouse)

This table lists the type of service and major suppliers. See Table 2.2 for full name and details of suppliers. Note, suppliers are often not the actual originators of the resources. All suppliers provide links to or information on the resources provided.

(PCR) pools. When filters are available they are often double spotted, which doubles the number of filters required but greatly reduces the number of false positives obtained. When assaying via PCR, the library is divided into a number of complex pools. Those testing positive are subdivided until the user arrives at a single clone. This typically takes two to four rounds of PCR depending on pool complexity. In the experience of the authors, if a choice of assay is available then PCR is usually the first choice. PCR is fast and easily scaleable, lending itself to automation at all stages. Hybridization screening has the advantage of allowing

Table 2.2 Details of suppliers

Name and location of supplier	Internet address
ATCC American Type Culture Collection, USA	www.atcc.org
Bios Bios, Yale, USA	mggsunc.ctan.yale.edu/home.html
CCR Corriel Cell Repositories, Camden, USA	arginine.umdnj.edu/ccr.html
CSM Cedar-Sinai Molecular Genetics	www.csmc.edu/genetics/korenberg/ reagents.page.html
CEPH (Fondation Jean Dausset), Paris, France	www.cephb.fr
ECACC European Collection of Animal Cell Cultures, Salisbury	www.gdb.org/annex/ecacc/HTML/ecacc.html
Genethon Genethon, Every, France	www.genethon.fr
Gen Sys Genome Systems, USA	www.genome.systems
Jackson Jackson Laboratories, Bar Harbor, USA	www.jax.org
JCRB Japanese Collection of Research Bioresources, Tokyo, Japan	www.nihs.go.jp
RP Roswell Park, Buffalo, USA	bacpac.med.buffalo.edu
Res Gen Research Genetics, Huntsville, USA	www.resgen.com
RIKEN Institute of Physical and Chemical Research, Saitama, Japan	www.riken.go.jp
RZPD German Human Genome Project, Berlin, Germany	www.rzpd.de
HGMP-RC United Kingdom Human Genome Mapping Project Resource Centre, Hinxton, UK	www.hgmp.mrc.ac.uk

The names and addresses of organizations referred to in Tables 2.1a and 2.1b are shown.
Note: All WWW addresses are preceded by http:// (the default for most browsers).

the use of anonymous probes, but a higher false-positive rate than for PCR screening is often seen. This necessitates further work to confirm the true positives.

When deciding which library to use, researchers will need to look at the average insert size, degree of chimerism (see below), coverage of the library (varies between libraries from 1x to 10x) and vector properties. These attributes will have differing effects depending on experimental needs.

2.1.1   Yeast Artificial Chromosome Libraries

The first practical system of presenting libraries. The yeast artificial chromosome (YAC) has the advantage of large inserts (200 kb to 2 Mb). This makes them ideal for isolating a region of interest (especially when available markers are sparse) and many YAC clones have been well characterized. The availability of the Centre d'Etude Polymorphisme Humain (CEPH) maps and their publication has maintained the high profile of the CEPH library (www.cephb.fr/ cephgenethon-map.html).

The biggest drawback in using YACs is the presence of appreciable levels of chimerism, meaning that two non-contiguous regions of DNA are presented in a contiguous manner (values of 10–60% of clones have been reported for various libraries). If an individual YAC is determined to be of interest to a researcher, it is usually necessary to confirm the integrity of the clone. Even so, the sheer size of the insert carried by YAC libraries ensures they remain popular, particularly in physical mapping programs.

2.1.2   Bacterial Artificial Chromosome Libraries

The major benefit of bacterial artificial chromosome (BAC) libraries is that they use plasmid-based vectors, removing many of the handling problems encountered with YACs. BACs have an average insert size of approximately 130 kb, requiring far more clones to cover the genome than with a YAC library. Chimerism is not a problem with BACs, and along with P1 Artificial Chromosome (PAC) libraries these are gaining in usage in sequencing projects. Because of the low copy number of the vector, large-volume purifications are required to obtain purified BAC DNA. Information on BACs is available (http://www.tree.caltech.edu/#30).

2.1.3   P1 Artificial Chromosome Libraries

Based on the pCYPAC and pPAC4 vectors, P1 artificial chromosome (PAC) libraries have an average 80–135-kb insert. No significant problems with chimerism have been reported. Many sequencing projects are now turning away from YACs and using PACs or BACs as their source material. Information on PACs is available (http://bacpac.med.buffalo.edu).

2.1.4   Cosmid Libraries

Cosmid libraries are based on *Escherichia coli*, contain an average insert of 40 kb and many use the Lawrist series of vectors. These utilize the *cos* packaging sites to generate the vectors. No problems with chimerism have been reported. Information on cosmids is available (www.bio-llnl.gov/genome/html/cosmid.html).

3   CHROMOSOME-SPECIFIC LIBRARIES

As more information about the genome becomes available, many researchers no longer need to screen the whole genome. Single chromosome libraries are now becoming available; most of the vectors have already been covered above.

3.1   Phage Libraries

A number of phage libraries covering most of the chromosomes are available, with an average insert size of about 20 kb.

4   CpG ISLAND LIBRARIES

The human genome contains approximately one-fifth of the statistically expected CpG content. It is possible to isolate regions where unmethylated CpGs occur at the expected rate. These CpG islands are found at the 5' end of many genes. It is estimated that about 60% of human genes and 50% of mouse genes are preceded by CpG islands. The value of CpG libraries is that they give access to the 5' end of genes while many cDNA libraries are biased towards the 3' end.

5   cDNA LIBRARIES

A wide variety of cDNA resources is available, falling into two broad categories: tissue, organ or developmentally specific libraries, and the IMAGE set.

5.1   Specific Libraries

These libraries have been generated from a wide variety of different sources and are presented in a variety of hosts and vectors. Most are generated from specific tissues or organs, although some have been generated from particular developmental stages. Some are provided as mixed libraries, whereas others are available as gridded filters. It should be noted that many of the clones from these various libraries have been made available via the IMAGE consortium (see below).

5.2   Integrated Molecular Analysis of Gene Expression (IMAGE) Consortium Library

The IMAGE consortium (www-bio.llnl.gov/bbrp/image/image.html) is a collaborative effort to produce high-quality cDNA libraries. The consortium shares cDNA libraries, placing all the sequence, map and expression data in the public databases (primarily dbEST). The individual clones are all assigned a unique IMAGE accession number. When a researcher identifies a clone of interest, it can then be ordered from selected suppliers. Some of the suppliers also make IMAGE clones available on filters. The number of clones in the library is continually increasing as more libraries are added.

6   PROBE AND PRIMER BANKS

Probe banks provide access to hybridization probes. Probe DNA is supplied as cloned DNA in a variety of vectors from which it must be purified. With the gradual move away from hybridization to PCR assays, primer banks have become available. As the cost of primer synthesis has fallen, the provision of primers has moved towards providing panels of primers, for example mapping panels, consisting of primers with specific chromosomal assignments.

7 CELL LINES

Human and model organism cell lines are available from a wide range of suppliers of which, by necessity, only a few will be highlighted here. These cell lines are provided to include a variety of mutants and recombinant materials.

8 CELL HYBRID PANELS

8.1 Somatic Cell Hybrid Panels

Somatic cell hybrid panels are the product of fusing the cell lines of two species. Human somatic cell panels are produced by fusing human and rodent cells. After several cycles of cell division, the chromosomes from the donor species are preferentially lost until only a small proportion remains. In the ideal circumstances only one chromosome is left. These individual cell lines are then made available as a mapping resource.

8.2 Radiation Hybrids

Radiation hybrids are produced in a similar manner to somatic cell hybrids, except that the donor chromosomes are irradiated and broken into fragments before fusion. Compared with monochromosomal panels, this has the additional advantage of of assigning a specific chromosomal location, via programs such as RHMapper.

9 BACKCROSS PANELS

A resource specific to the mouse, backcross panels allow users to map their markers against an interspecific backcross. This has the benefit of producing integrated maps and removes the need to carry out a backcross for a simple mapping experiment. Two available panels result from 188 (Jackson) and 982 (EUCIB) mice backcrosses.

3 Managing Pedigree and Genotype Data

Stephen P. Bryant

Gemini Research Limited, 162 Science Park, Milton Road, Cambridge CB4 4GH, UK
E-mail: Steve.Bryant@gemini-research.co.uk

1 INTRODUCTION

Model-based or model-free analyses of genetic linkage data remain popular ways of mapping disease genes using extended families or affected sib-pairs, and the statistical ramifications of this work are well developed. A bottleneck in any mapping project remains the management of information to facilitate the process, particularly with regard to the transformation of data formats and support for statistical analysis. Previous reviews of the subject (Bryant, 1991, 1994, 1996) have concentrated on available software solutions, or have considered research and development projects that have showed great promise, such as the Integrated Genomic Database (IGD) (Bryant *et al.*, 1997). Very little has changed over the last two years to alter the basic message that these reviews provide, and I do not attempt to cover the same ground here.

In this chapter, I take the domain of human pedigree analysis, with particular reference to the management of genotypes and phenotypes, and consider how experimental data generated by this kind of work can be managed effectively. The most crucial observation is that there are currently no Windows, Macintosh, Unix or DOS applications that can be recommended as being sufficiently complete to serve the needs of the contemporary mapping laboratory, which may be performing genome scans on large collections of sib-pairs, or more selective typing across extended families. Some specialized systems are available and are very good, but they do not embed well within an existing laboratory setup without significant modifications. What I have created in this treatment, however, is something fairly novel for the molecular biologist to use as a communication tool with the bioinformatician, or for bioinformaticians to use as a roadmap to guide them through the creation of a customized system for handling these kinds of data.

I base the treatment around well established data modelling techniques such as Entity Relationship Attribute (ERA) analysis (Chen, 1976), which is an appropriate

Guide to Human Genome Computing, 2nd edition   Copyright © 1998 Academic Press Limited
ISBN 0-12-102051-7   All rights of reproduction in any form reserved

design methodology for relational database systems, and provide example code in Structured Query Language (SQL) form. The intention is to provide a basis for the construction of a real working system, which can be implemented in one of a number of industrial database technologies, such as Oracle, Sybase or Informix, as well as more leading-edge systems, such as ObjectStore. The bioinformatician can use the guide as a recipe book, from which to pick and choose as appropriate.

2 FUNDAMENTALS

Consider the principles of a model-based analysis. An *analysis* is an application of a *model* to a set of *observations*. The observations are often a set of phenotypes and genotypes recorded on a group of related individuals, termed a *pedigree* or *genealogy*. For a linkage analysis between a trait, influenced by the genotype at a susceptibility locus, and a polymorphism suspected to be nearby on the same chromosome, the model is a combination of:

1. The inheritance model of the trait;
2. The inheritance model, allele and genotype frequency distribution of the linked polymorphism; and
3. The recombination fraction(s) between the trait and linked polymorphism, which may be a single parameter (θ), averaging the effects of differences between the sexes, or two parameters (θ_m and θ_f), defining the sex-specific recombination fractions where θ_m and θ_f are estimated separately.

The following sections consider this in more detail.

Figure 3.1 *An example pedigree, shown in conventional graphical form. Each square or circle represents a person. Squares represent males and circles females. The unique identifier for each person is given below each symbol. A diagonal line through the symbol means that the person is known to be dead. A vertical line is used to connect parents with offspring. A horizontal line is used to join sibships and also to join parents. A trait T is segregating in the family, inherited as an autosomal dominant, fully penetrant trait with two phenotypes: (1) affected (shown as a black filled symbol), and (2) unaffected (shown as a white-filled symbol). People with unknown T phenotype contain an interrogation mark (?). Codominant marker loci A, B, C, D and E (from top of diagram to bottom) are also shown. Genotypes are given as pairs of alleles A_iA_j, B_iB_j, C_iC_j, D_iD_j or E_iE_j. Unknown genotypes are scored as 0 0. Paternal alleles are shown on the left of each genotype pair, maternal on the right. Alleles unambiguously inherited from the same grandparent are shown boxed, as black (grandpaternal) and white (grandmaternal). Crossovers are shown as a change of phase from grandpaternal to grandmaternal, or vice versa.*

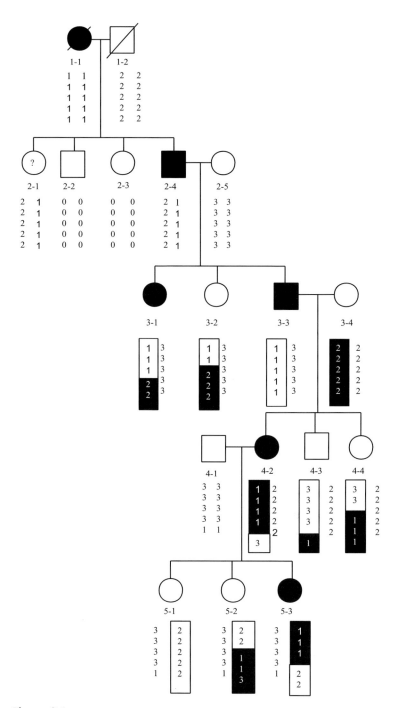

Figure 3.1

2.1   Pedigrees

The pedigree (Figure 3.1) is the unit of information in a linkage analysis. The observations on a pedigree have a special kind of interdependency, determined by the mechanism of genetic inheritance. Human pedigrees or genealogies can be classified into several kinds on the basis of the distribution of *affecteds* (i.e. individuals expressing a particular trait) (Table 3.1).

For simple family types such as sib-pairs or nuclear three-generational families, there typically exist a large number of identical family structures. For more complicated family types (Figure 3.1), there is typically only one family of a given structure and distribution of affecteds in any one study. These differences influence how the information from different families can be combined, but do not impact on how the data are stored or managed.

2.2   Traits and Trait Models

In general, a trait can be defined as any phenotypic characteristic. Typically, this may be the presence or absence of a morbid character, such as a disease. In this case, the trait is qualitative and individuals are scored with respect to their *affection status*, and are either *affected* or *unaffected*. Traits may also be quantitative, where the phenotype takes a value from a continuum. These include height, weight, blood pressure and levels of blood sugar.

How can these phenotypes be related to an underlying genotype? If the trait, denoted by T, can be considered largely the result of genotypic variation at a single locus, then the trait can be modelled as a two-allele system (t and n) with frequencies p_t and $p_n = (1 - p_t)$, in combination with several penetrance parameters, which define the probability of expressing the trait phenotype conditional on having a particular genotype (tt, tn or nn for autosomal traits, t or n for hemizygous males and X-linked traits) at the trait locus (Table 3.2).

This method of defining the genetics of a trait is generally applicable (Table 3.3) but will have reduced power (due to reduced penetrance and increased phenocopy rate) when the true underlying model is polygenic or

Table 3.1   Classification of family types, by distribution of affecteds

Family type	Description
Simplex	Single affected offspring, with parental information
Sib-pairs	Pairs of affected siblings, with or without parental information
Sib-sets	As above, with more than two offspring
Relative pairs	Pairs of relatives in relationships other than sibs, with or without information from other family members
Extended	General case, multiplex families in arbitrary degrees of relationship

Table 3.2 Modelling the expression of a trait phenotype

Parameter	Meaning
p_t	Trait allele frequency = $1 - p_n$
f_{tt}	Penetrance of the t/t genotype = $p(T|tt)$
f_{tn}	Penetrance of the t/n genotype = $p(T|tn)$
f_{nn}	Penetrance of the n/n genotype = $p(T|nn)$
f_t	Penetrance of the t allele = $p(T|t)$
f_n	Penetrance of the n allele = $p(T|n)$

Table 3.3 A selection of qualitative trait models, showing how varying the penetrance parameters can model the segregation of the phenotype (parameters not used in the model are indicated by a dash)

Name	p_t	f_{tt}	f_{tn}	f_{nn}	f_t	f_n	Examples
Fully penetrant autosomal dominant	0.001	1.0	1.0	0.0	—	—	Adenomatous polyposis coli (Bodmer *et al.*, 1987); non-epidermolytic palmoplantar keratoderma (Kelsell *et al.*, 1995)
Fully penetrant autosomal recessive	0.04	1.0	0.0	0.0	—	—	Muscular dystrophy with epidermolysis bullosa (Smith *et al.*, 1996)
Fully penetrant X-linked recessive	0.04	1.0	0.0	0.0	1.0	0.0	Charcot–Marie–Tooth neuropathy (Erwin, 1944)
Partially penetrant autosomal dominant	0.003	0.4	0.4	0.02	—	—	Early-onset breast cancer (Williams and Anderson, 1984)

embodies some other variation. In this case, a better fit to the model can be obtained by considering additional parameters, such as the possibility of modifying loci, although these are not considered further here.

2.3 Markers and Marker Models

Linkage analysis has been facilitated greatly by the discovery of highly polymorphic regions of the human genome, revealed with the techniques of modern molecular biology. Di-, tri- and tetra-nucleotide repeats and the use of polymerase chain reaction (PCR) have all contributed to the availability of highly polymorphic markers, and microsatellites now predominate in maps of the human

Table 3.4   Modelling the marker system

Parameter	Meaning
$a_1, ..., a_n$	Marker allele frequencies $(0 < a_n \leq 1.0)$
$a_1a_1, ..., a_na_n$	Genotype frequencies

genome (Buetow *et al.*, 1994; Gyapay *et al.*, 1994) and are the most usual choice of polymorphism for mapping work. Markers are typically codominant and can be modelled as a set of population allele and genotype frequencies (Table 3.4).

2.4   Phenotypes, Genotypes and Haplotypes

Qualitative (or affection status) phenotypes may be recorded as a binary presence/absence tag, typically *U* (Unaffected) and *A* (Affected). Quantitative phenotypes can be recorded as a Real number.

Autosomal genotypes can be recorded as a pair of unordered integers (1/1, 1/2, 2/2, etc.) after being recoded from allele length, fragment composition or other mechanism. X-linked genotypes may be recorded as a pair of unordered integers (females) or as a single integer (males). Figure 3.1 shows how haplotypes may be recorded explicitly on a figure, where alleles inherited from the same parent and therefore residing on the same chromosome are shown together on the same side of each genotype, with paternal chromosomes on the left and maternal on the right. The shading represents the *grandparental* source of each allele. This representation makes it immediately evident where genetic recombination has occurred, where the grandparental origin of an allele changes from paternal to maternal, or vice versa.

2.5   The Analysis

The linkage analysis is performed by applying the model (developed in Sections 2.2 and 2.3) to a set of observations (Sections 2.1 and 2.4). A subset of pedigrees, markers and trait phenotypes (typically a single phenotype) is selected and the data are transformed into the format required for analysis, which will typically include the genetic model of the trait and marker(s). Usually, the analytical software will be one of LINKAGE (Lathrop and Lalouel, 1984), GAS (A. Young, University of Oxford, 1993–1995; available from http://users.ox.ac.uk/~ayoung/gas.html), GeneHunter (Kruglyak *et al.*, 1996), CRI-MAP (Lander and Green, 1987) or similar, rather than a general-purpose statistical package, such as SPSS or STATA. It might also be useful to display the pedigree, genotypes and phenotypes graphically, in a way similar to Figure

3.1, using a system such as PEDRAW (Curtis, 1990) or Pedigree/Draw (Dyke and Mamelka, 1987).

Some work has been done on more extensive integration of the analytical process (Bryant *et al.*, 1997), although more usually results are not typically imported back into the database, but are instead managed separately, as part of the general file system.

3 MODELLING THE DOMAIN

The process of Entity Relationship Attribute (ERA) analysis was first described by Chen (1976), and has since been subject to many modifications and extensions. It is a process whereby a domain is formalized into a set of objects (entities) connected by relationships. Attributes are associated with either entities or relationships. The distinction between an entity and a relationship is not always informative, and it is perfectly satisfactory to model the domain in terms of entities alone. In the discussion that follows, any object of interest will be defined as an entity. An ERA model is a graphical structure, used as an early design step as an aid to communication between the bioinformatician and the pedigree analyst or molecular biologist. It is a useful although time-consuming part of the design process, providing a model that can be translated rapidly into working code. The internal consistency of such a model can also be verified easily.

There are several ways in which we can break the domain down into entity units. There is no single correct way of representing a real-world situation, such as the flow of genes through a pedigree, with this kind of model. Models are more or less expressive, more or less useful. Their utility will be evident by trying out example data.

From the basic treatment, developed in the previous section, a data model can be developed that forms the basis of a working system. The first consideration is the modelling of the:

1. Pedigree structure
2. Trait and trait model
3. Marker and marker model
4. Phenotypes, genotypes and haplotypes
5. Analysis.

3.1 The Pedigree Structure

In any genetic analysis, the basic information unit is the pedigree. All pedigree types (Table 3.1) can be modelled using the same basic method, irrespective of the simplicity or complexity of the structure (Figure 3.2).

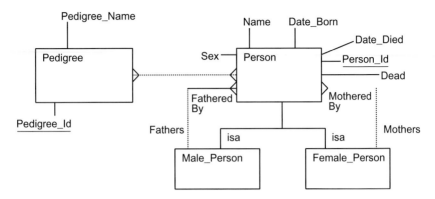

Figure 3.2 *Entity modelling of pedigree structure. Entities are shown as rectangular, with related entities connected by solid or dashed lines. The cardinality of the relationship (1 : 1, 1 : N or M : N) is shown by crows-foot notation, and the optionality of the relationship (mandatory or optional) is given by solid or dashed lines respectively. Selected attributes are shown for each entity. Attributes forming the primary key are shown underlined.*

Figure 3.2 illustrates several diagrammatic conventions. Entities are shown as rectangular boxes. Each entity has a primary key, which is that attribute or combination of attributes that together uniquely define an instance of that entity. Attributes forming the primary key are shown underscored. Relationships between entities are shown as solid or dashed lines. If an instance of an entity must take part in a relationship with an instance of another entity (*obligate* or *mandatory* inclusion), the end of the line forming the relationship is solid. If participation is optional, the line is dashed. Therefore, the relationship between Pedigree and Person should be read as:

'A pedigree may contain one or more people'

and

'A person may belong to one or more pedigrees'

Families are typically identified by a numeric or alphanumeric identifier (`Pedigree Id`) and Individuals by their own identifier (`Person_Id`). Individuals can be assigned to pedigrees by participating in the relationship between `Pedigree` and `Person`.

Another interesting feature of such a structure is that pointers to parents need to be provided. One way to model this is to include references, within the `Person` entity, to other instances of `Person` representing the `Father` and `Mother` (Figure 3.2). Since the `Person` entity divides neatly into the mutually exclusive subtypes `Male_Person` and `Female_Person` (i.e. a person cannot be a member of both the `Male_Person` and `Female_Person` subtypes at the same time), these relationships can be expressed more completely by the subtypes (`Male_Person` and `Female_Person`) participating in the relationship, rather than the supertype (`Person`).

This illustrates the notion of roles in a relationship. These are made explicit in the figure by showing `Male_Person` and `Female_Person` in an optional role of Father (or Mother) to the entity `Person`, i.e. only males may be fathers (to one or more people). There is a small element of redundancy in that being a father implies being male, with the sex of a person also explicitly recorded elsewhere. Redundancy is permissible within a model, and requires careful handling of transactions on an implemented database, to ensure integrity is maintained. A selection of suggested useful attributes is also provided on the figure.

From these basic beginnings, there are a lot of biological constraints that have not been captured in the model as it stands, and which can be added as 'arbitrary' constraints. A more formal modelling language, such as VDM or Z, could capture these constraints in a more unified way. Constraint handling is considered in more detail in Section 4.

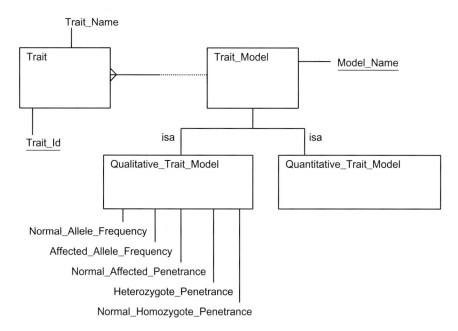

Figure 3.3 *Entity modelling of traits and trait models. For qualitative traits that are influenced largely by the genetics at a single locus, the model provides a two-allele system (t and n), with penetrances for the tt, tn and nn genotypes. This is sufficient to express the genetics of all the traits in Table 3.4, and many more besides.*

3.2   The Trait and Trait Model

The way in which the expression of a phenotype is influenced by an underlying genotype can be modelled as a single entity (Figure 3.3). Traits may be identified by a `Trait_Id`, with an alternative short name (`Trait_Name`) as a candidate key. Considering phenotypes influenced by the genotype at a single underlying locus, such as those in Table 3.4, the trait model used is simply expressed as a combination of allele frequencies and penetrances of the trait (T) and normal (n) alleles, as shown.

3.3   The Marker and Marker Model

Marker systems can be modelled simply as a set of allele frequencies (Figure 3.4). In a stable homogeneous population, genotype frequencies will be simple function of their component allele frequencies. It may, in some cases, be of value to be able to specify genotype frequencies, which can

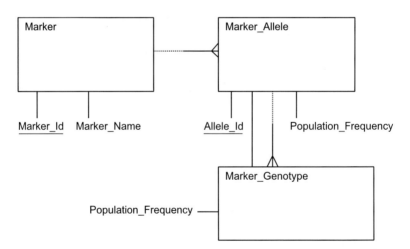

Figure 3.4 *Entity modelling of markers and marker models. For many markers, the model is simply one representing the frequency of each allele in the population from which the families were drawn. A more complex model could consider the idea of population, and include another entity to allow allele frequencies to vary in an analysis. For most marker systems and most populations, the population genotype frequencies will be a simple function of allele frequencies. Allele labels are provided to permit comparisons to be made across datasets. To combine different datasets in an analysis, it is crucial that the recoding done to convert, for example, allele lengths into recoded alleles is the same for each population. The Genome Data Base (GDB: http://gdbwww.gdb.org) is the authoritative source for this information.*

be represented as a relationship between alleles with the additional attribute `Population_Frequency` (Figure 3.4).

3.4 Phenotypes and Genotypes

An observed genotype is an instance of a relationship between `Person` and alleles at a genetic locus, of which the observation is one of many possibilities. Figure 3.5 shows how the graphical representation of Figure 3.1 can be applied to extending the model in Figure 3.2, to include information on genotype and phenotype. Note the reference to entities (`Qualititative_Trait` and `Marker_Genotype`) that have been described already. The distinction between autosomal and X-linked genotypes is also made. The extension to cover haplotypes is omitted.

3.5 The Analysis

The recommended way of interoperating with analytical software is to provide (i) the means of selecting a subset of information for analysis, and (ii) an algorithm for generating the input formats required by the program. Case (i) can be supported by providing a model such as that shown in Figure 3.6, which introduces the concepts of a `Pedigree_Set` and `Marker_Set` which can be combined with a `Trait` and `Trait_Model` into an `Analysis`, to be reused as and when required. Case (ii) can be achieved by creating the SQL code to generate the layout from the chosen `Pedigree_Set`, `Marker_Set`, `Trait` and `Trait_Model`, and is covered in Section 4.

4 IMPLEMENTING THE DATABASE

From the abstract structures developed in Section 3, an implementation can be created by mapping on to the relational model, followed by a further mapping to SQL, and this is considered in this section.

4.1 Mapping on to the Relational Model

The mapping of the ERA model into relations is straightforward. As a first step, all entities generate their own relation. For *1 : n* relationships, much of the process is concerned with posting the primary key of the entity on the *1* side of the

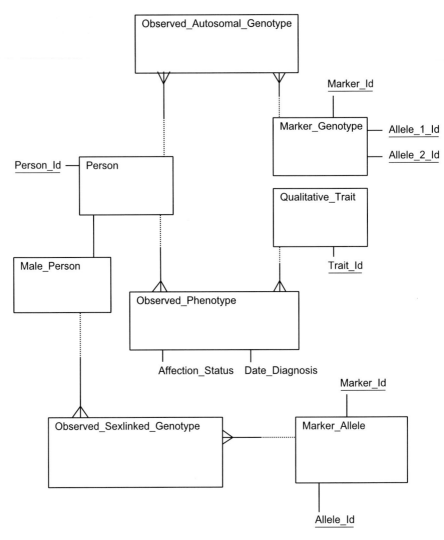

Figure 3.5 *Entity modelling of phenotypes and genotypes. This model represents one observed genotype at each marker locus for each person, and one observed phenotype (trait measurement) on each trait for each person.*

relationship into the entity on the *n* side of the relationship, where the attribute set becomes a foreign key and points to a primary key in the 1 entity, by value. These correspondences can be used to construct query results consisting of attributes from more than one table.

The process of normalization is necessary to eliminate attribute dependencies on anything other than the primary key. Dependencies lead to redundancy and risk compromising the database through inconsistency. Some designs may be well normalized at an early stage. Otherwise, relations can be decomposed to remove any repeating groups.

Subtypes (e.g. `Qualitative_Trait` and `Quantitative_Trait`) have already been mentioned briefly. There are several ways to implement subtypes within the relational model. One is to create an explicit relation for the main type (supertype) and one for each subtype. This is useful if there are attributes that are specific to one subtype but meaningless for another. Alternatively, the

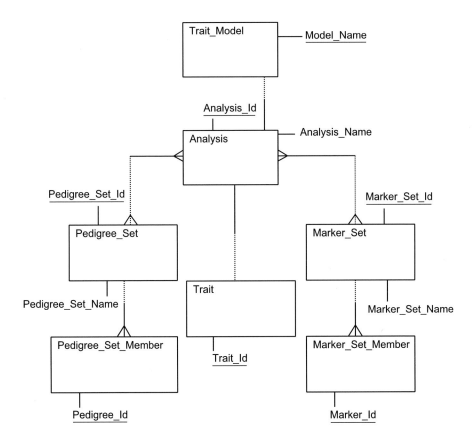

Figure 3.6   *Entity modelling of the analysis. Analyses can be supported by the creation of Pedigree and Marker sets, linked by an Analysis entity, facilitating reuse. The Pedigree Set and Marker Set entities have a many-to-many (M : N) relationship with Analysis, which means that one Pedigree Set can be used by many Analyses, and that Analysis can be constructed from many Pedigree Sets. This is a very expressive way of selecting a subset of data. The analysis is also associated with a single* `Trait` *and a single* `Trait_Model`.

```
Create table pedigree (pedigree_id number not null,
pedigree_name varchar2(20) null);
Alter table pedigree add primary key (pedigree_id);
Create table person (person_id number not null, name
varchar2(30) not null, date_born date null, date_died
date null, sex varchar2(1) not null, dead varchar2(1));
Alter table pedigree add primary key (pedigree_id);
Create table pedigree_person (pedigree_id number not
null, person_id number not null);
Alter table pedigree_person add primary key
(pedigree_id, person_id);
Create view male_person as select person_id from person
where sex = 'M';
Create view female_person as select person_id from
person where sex = 'F';
Create table father (person_id number not null,
father_id number not null);
Alter able father add primary key (person_id,
father_id);
Create table mother (person_id number not null,
mother_id number not null);
Alter table mother add primary key (person_id,
mother_id);
Alter table person add constraint check_sex check (sex
in ('M', 'F', 'U'));
```

Figure 3.7   *SQL representation of Figure 3.2 (the pedigree structure).*

```
Create table Trait (Trait_Id number not null, Trait_Name
varchar2(20) null);
Alter table Trait add primary key (Trait_Id);
Create table Trait_Model (Model_Name varchar2(30) not
null, Model_Type varchar2(3) not null,
normal_allele_frequency float, affected_allele_frequency
float, normal_affected_penetrance float,
heterozygote_penetrance float,
normal_homozygote_penetrance float);
Alter table Trait_Model add primary key (Model_Name);
Create view Qualitative_Trait_Model as select
model_name, normal_allele_frequency,
Affected_allele_frequency, normal_affected_penetrance,
heterozygote_penetrance,
Normal_homozygote_penetrance from Trait_Model where
model_type = 'QLT';
Create view quantitative_trait_model as select
model_name from trait_model where model_type = 'QNT';
```

Figure 3.8   *SQL representation of Figure 3.3 (the trait and trait model).*

subtypes could be views of the supertype, using an attribute flag or other restriction criterion for obtaining the members of each subtype. This is the easiest solution to implement. Examples are given in Figures 3.7–3.11.

The database can be implemented in any one of a number of flavours of SQL. The process of mapping relations on to a Data Description Language (SQL DDL) is largely automatic. Each table is created using the CREATE TABLE statement in SQL. Primary key, foreign key and arbitrary constraints are set up in a slightly different way by each variant of SQL. The examples given in Figures 3.7–3.11 are all for the Oracle dialect and could be easily adapted to Sybase, SQL server or similar technology by adjusting the syntax. Examples of biologically meaningful constraints are provided.

```
Create table marker (marker_id number not null,
marker_name varchar2(20) null);
Alter table marker add primary key (marker_id);
Create table marker_allele (marker_id number not null,
allele_id number not null, allele_name varchar2(4) null,
population_frequency float null);
Alter table marker_allele add primary key (marker_id,
allele_id);
Create table marker_genotype (marker_id number not null,
allele_1_id number not null, allele_2_id number not
null, population_frequency float null);
Alter table marker_genotype add primary key (marker_id,
allele_1_id, allele_2_id);
Alter table marker_genotype add constraint allele_order
check (allele_1_id ≤ allele_2_id);
```

Figure 3.9 *SQL representation of Figure 3.4 (the marker and marker model).*

```
Create table observed_autosomal_genotype (person_id
number not null, marker_id number not null, allele_1_id
number not null, allele_2_id number not null);
Alter table observed_autosomal_genotype add primary key
(person_id, marker_id);
Create table observed_phenotype (person_id number not
null, trait_id number not null, affection_status
varchar2(1) not null, date_diagnosis date null);
Alter table observed_phenotype add primary key
(person_id, trait_id);
Create table observed_sexlinked_genotype (person_id
number not null, marker_id number not null, allele_id
number not null);
```

Figure 3.10 *SQL representation of Figure 3.5 (phenotypes and genotypes).*

```
Create table analysis (analysis_id number not null,
analysis_name varchar2(30), model_name varchar2(30) not
null, trait_id number not null);
Alter table analysis add primary key (analysis_id);
Create table analysis_pedigree_set (analysis_id number
not null, pedigree_set_id number not null);
Alter table analysis_pedigree_set add primary key
(analysis_id, pedigree_set_id);
Create table analysis_marker_set (analysis_id number not
null, marker_set_id number not null);
Alter table analysis_marker_set add primary key
(analysis_id, marker_set_id);
Create table pedigree_set (pedigree_set_id number not
null, pedigree_set_name varchar2(10) null);
Alter table pedigree_set add primary key
(pedigree_set_id);
Create table pedigree_set_member (pedigree_set_id number
not null, pedigree_id number not null);
Alter table pedigree_set_member add primary key
(pedigree_set_id, pedigree_id);
Create table marker_set (marker_set_id number not null);
Alter table marker_set add primary key (marker_set_id);
Create table marker_set_member (marker_set_id number not
null, marker_id not null);
Alter table marker_set_member add primary key
(marker_set_id, marker_id);
```

Figure 3.11   *SQL representation of Figure 3.6 (the analysis).*

5   INTERFACING WITH ANALYSIS PROGRAMS

Rather than model the analysis completely, the examples provided have modelled the selection of a pedigree group and a marker group. The most important function is the efficient generation of the format for analysis. As an example, the generation of data for LINKAGE is provided as an annotated code fragment (Figure 3.12). In this figure, and also in Figure 3.13, the PL/SQL code is embedded inside a csh script, which can be executed on most Unix systems. It includes a call to the Oracle SQL interpreter, sqlplus, which places output lines into a table called Results, which is then output using a standard SQL select statement. Lines are ordered by including the row number in each line, and the possibility of competing processes interfering with each other (if Results is a shared table) is removed by including the Unix process id ($$) as a tag for each row in the table. The Results table is cleaned up at the end of the script.

```
#!/bin/csh -f
#
# This script converts ORACLE data into LINKAGE format. The data are written
# to a pair of files, following LINKAGE convention.
#
# Usage : ora2lkg <orauser> <orapwd> <analysis>
# where <analysis> is the analysis_name of the analysis, associated with a
# pedigree_set, marker_set, trait and trait_model.
#
setenv ORA_HOME /home/oracle
setenv SQL $ORA_HOME/bin/sqlplus
setenv ORAUSER $1
setenv ORAPWD $2
setenv ANALYSIS $3

# First, create the pedigree file of genotypes and phenotypes
$SQL -s $ORAUSER/$ORAPWD <<END > $ANALYSIS.ped
define analysis='$ANALYSIS'

declare
  /* set up the structure to hold all the basic analysis information */
  cursor analysis_cursor is select analysis_id, analysis_name,
  trait_id, model_name from analysis where analysis_id = &analysis;
  analysis_record analysis_cursor%rowtype;

  cursor pedigree_cursor is select pedigree_id, pedigree_name
  from pedigree where pedigree_id in (select pedigree_id from pedigree_set
  where pedigree_set_id in (select pedigree_set_id from
  analysis_pedigree_set where analysis_id = &analysis)) order by pedigree_id;
  pedigree_record pedigree_cursor%rowtype;

  cursor person_cursor is select person_id, sex, name from person
  where person_id in (select person_id from pedigree_person where
  pedigree_id = pedigree_record.pedigree_id) order by person_id;
  person_record person_cursor%rowtype;

  cursor marker_cursor is select marker_id, marker_name from marker
  where marker_id in (select marker_id from marker_set_member
  where marker_set_id in (select marker_set_id from
  analysis_marker_set where analysis_id = &analysis)) order by marker_id;
  marker_record marker_cursor%rowtype;

  cursor genotype_cursor is select allele_1_id, allele_2_id
  from observed_autosomal_genotype where marker_id = marker_record.marker_id
  and person_id = person_record.person_id;
  genotype_record genotype_cursor%rowtype;

  cursor phenotype_cursor is select affection_status from observed_phenotype
  where trait_id = analysis_record.trait_id and person_id =
person_record.person_id;
  phenotype_record phenotype_cursor%rowtype;
```

Figure 3.12 Continued

```
    cursor father_cursor is select father_id from father
    where person_id = person_record.person_id;
    father_record father_cursor%rowtype;

    cursor mother_cursor is select mother_id from mother where
    person_id = person_record.person_id;
    mother_record mother_cursor%rowtype;

    result varchar2(2000); rowseq number(8);

begin
    rowseq:=1; open analysis_cursor; fetch analysis_cursor into analysis_record;
    close analysis_cursor;
    open pedigree_cursor; loop fetch pedigree_cursor into pedigree_record;
      exit when pedigree_cursor%notfound;
      result:= rpad(to_char(pedigree_rec.pedigree_id),8);
      open person_cursor; loop fetch person_cursor into person_record;
        exit when person_cursor%notfound;
        result := result || rpad(to_char(person_record.person_id),8);
        open father_cursor; fetch father_cursor into father_record;
        if father_cursor%notfound then result := result || rpad('0',8);
        else result := result || rpad(to_char(father_record.father_id),8);
        end if; close father_cursor;
        open mother_cursor; fetch mother_cursor into mother_record;
        if mother_cursor%notfound then result := result || rpad('0',8);
        else result := result || rpad(to_char(mother_rec.mother_id),8);
        end if; close mother_cursor;
        result := result || rpad(person_record.sex,2);
        open phenotype_cursor; fetch phenotype_cursor into phenotype_record;
        if phenotype_cursor%notfound then result := result || rpad('0',8);
        else result := result || rpad(to_char(aff_rec.affection_status,3));
        end if; close phenotype_cursor;
          open marker_cursor; loop fetch marker_cursor into marker_record;
            exit when marker_cursor%notfound;
            open genotype_cursor; fetch genotype_cursor into genotype_record;
            if genotype_cursor%notfound then result := result || rpad(
                '0',3) || ' ' || lpad('0',3) ;
            else result := result || rpad(to_char(genotype_record.p_allele),3)
            || ' ' || lpad(to_char(genotype_record.m_allele),3);
            end if; close genotype_cursor;
          end loop; close marker_cursor;
        end loop;
      insert into results(result,rowseq,process_id) values (result,rowseq,&pid);
      rowseq:=rowseq+1;
    end loop; close pedigree_cursor;
end;
/

/* Generate the output file as an ordered sequence of lines */
select result from results where process_id = &pid order by rowseq;
```

Figure 3.12   Continued

```
/* Now clean up the results table and make the changes permanent. */
delete from results where process_id = &pid; commit;

quit
END

# Now create the file of locus information

$SQL -s $ORAUSER/$ORAPWD <<END > $ANALYSIS.par
define analysis='$ANALYSIS'

declare
  cursor analysis_cursor is select analysis_id, analysis_name, trait_id,
model_name
    from analysis where analysis_id = &analysis;
  analysis_record analysis_cursor%rowtype;

  cursor trait_cursor is select trait_id, trait_name
  from trait where trait_id = analysis_record.trait_id;
  trait_record trait_cursor%rowtype;

  cursor trait_model_cursor is select normal_allele_frequency,
  affected_allele_frequency, normal_affected_penetrance,
  heterozygote_penetrance, affected_homozygote_penetrance
  from trait_model where model_name = analysis_record.model_name;
  trait_model_record trait_model_cursor%rowtype;

  cursor marker_cursor is select marker_id, marker_name from marker
  where marker_id in (select marker_id from marker_set_member
  where marker_set_id in (select marker_set_id from
  analysis_marker_set where analysis_id = &analysis)) order by marker_id;
  marker_record marker_cursor%rowtype;

  local_rowseq number(8); local_result varchar2(2000);
  local_marker_count number(8);

begin

  open analysis_cursor; fetch analysis_cursor into analysis_record;
  close analysis_cursor;

  /* LINE 1 */ local_marker_count := 0; local_rowseq:=1; local_result:='';

  select count(marker_id) into local_marker_count from marker_set_member
  where marker_set_id in (select marker_set_id from analysis_marker_set
  where analysis_id = &analysis);

  local_result := to_char(local_marker_count+1) || '   0   0   3';
  insert into results(result,rowseq) values (local_result,local_rowseq);
  local_rowseq:=local_rowseq+1;
```

Figure 3.12 Continued

```
/* LINE 2 */ local_result := '0    0.0    0.0    0';
insert into results(result,rowseq) values (local_result,local_rowseq);
local_rowseq:=local_rowseq+1;

/* LINE 3 */ i := 1; local_result := '';

loop local_result := local_result || to_char(i) || '    ';
    if (i=local_marker_count+1) then exit; end if; i := i + 1; end loop;

insert into results(result,rowseq) values (local_result,local_rowseq);
local_rowseq:=local_rowseq+1;

/* LINE 4 */ open trait_cursor; fetch trait_cursor into trait_record;
close trait_cursor;

local_result := '1    2    #' || trait_record.trait_name;
insert into results(result,rowseq) values (local_result,local_rowseq);
local_rowseq:=local_rowseq+1;

/* LINE 5 */ open trait_model_cursor; fetch trait_model_cursor into
trait_model_record; close trait_model_cursor;

local_result := trait_model_record.normal_allele_frequency
    || trait_model_record.affected_allele_frequency;
insert into results(result,rowseq) values (local_result,local_rowseq);
local_rowseq:=local_rowseq+1;

/* LINE 6 */ local_result := '1';
insert into results(result,rowseq) values (local_result,local_rowseq);
local_rowseq := local_rowseq+1;

/* LINE 7 */ local_result := trait_model_record.normal_homozygote_penetrance
||
    trait_model_record.heterozygote_penetrance ||
    trait_model_record.affected_homozygote_penetrance;
insert into results(result,rowseq) values (local_result,local_rowseq);
local_rowseq:=local_rowseq+1;

open marker_cursor; loop fetch marker_cursor into marker_record;
    exit when marker_cursor%notfound;

    /* LINE 8 */ local_allele_count := 0;

    select count(*) into local_allele_count from marker_allele
    where marker id = marker record.marker id;

    local_result := '3 ' || to_char(local_allele_count) || '    # ' ||
        marker_set_member_record.marker_id;
    insert into results(result,rowseq) values (local_result,local_rowseq);
    local_rowseq:=local_rowseq+1;
```

Figure 3.12 Continued

```
/* LINE 9 */ local_result := '';

open allele_cursor; loop fetch allele_cursor into allele_record;
  exit when allele_cursor%notfound;
  local_result := local_result || rpad(allele_record.frequency,8);
end loop; close allele_cursor;

local_result := local_result || ' ';
insert into results(result,rowseq) values (local_result,local_rowseq);
local_rowseq:=local_rowseq+1;
end loop; close marker_cursor;

/* LINE 10 */ local_result := '0    0';
insert into results(result,rowseq) values (local_result,local_rowseq);
local_rowseq:=local_rowseq+1;

/* LINE 11 */ open marker_cursor; local_result := '';

loop fetch marker_set_member_cursor into marker_set_member_record;
  exit when marker_set_member_cursor%notfound;
  local_result := local_result || '0.1 ';
end loop; close marker_set_member_cursor;

insert into results(result,rowseq) values (local_result,local_rowseq);
local_rowseq:=local_rowseq+1;

/* LINE 12 */ local_result := '1';
insert into results(result,rowseq) values (local_result,local_rowseq);
local_rowseq:=local_rowseq+1;

/* LINE 13 */ local_result := '0    0';
insert into results(result,rowseq) values (local_result, local_rowseq);
local_rowseq:=local_rowseq+1;

end;
/

/* Generate the output file as an ordered sequence of lines. */
select result from results where process_id = &pid order by rowseq;

/* Now clean up the results table. */
delete from results where process_id = &pid;

/* And make the changes permanent. This is faster than a rollback. */
commit;

quit
END
```

Figure 3.12 *SQL to generate files for LINKAGE, from the data model developed in Figures 3.2 – 3.11.*

```
#!/bin/csh -f
#
# This script converts ORACLE data into a basic
# Pedigree/DRAW format. The data is written to the
# standard output.
#
# Usage : ora2pedraw <orauser> <orapwd> <pedigree>
#
# where <pedigree> is the pedigree_name of the family.
#
#

setenv ORACLE_HOME /home/oracle
setenv ORAUSER $1
setenv ORAPWD $2
setenv PEDIGREE $3
sqlplus -s $ORAUSER/$ORAPWD <<END
set heading off
set feedback off
set pagesize 0
set linesize 300
set verify off
ttitle off
define pid=$$
declare

/*
|| set up the structure to hold the basic pedigree
|| information.
*/

cursor pedigree_cursor is
select p.pedigree_id, pedigree_name, to_char(sysdate,
'DD Mon YYYY')
from pedigree p
where p.pedigree_name = '$PEDIGREE';

pedigree_rec pedigree_cursor%rowtype;

/*
|| set up the structure to hold the list of people associated
|| with the pedigree
*/

cursor person_cursor is
select p.person_id,
```

Figure 3.13 Continued

```
nvl(f.father_id,0) father_id,
nvl(m.mother_id,0) mother_id,
p.name,
nvl(p.sex,'U') sex,
nvl(dead,'?') dead,
nvl(to_char(p.date_born,'Mon DD, YYYY'),'?') date_born,
nvl(to_char(p.date_died,'Mon DD, YYYY'),'?') date_died,
decode(dead,'Y',
decode(dead,'Y',floor(months_between(p.date_died,p.date_born)/
12),0) age_at_death,
decode(dead,'Y',floor(months_between(sysdate,p.date_born)/12),
0)current_age
from person p,pedigree_person pp,father f,mother m
where pp.pedigree_id = pedigree_rec.pedigree_id and
pp.person_id = p.person_id and
pp.person_id = f.person_id (+) and
pp.person_id = m.person_id (+) and
order by p.person_id;

person_rec person_cursor%rowtype;

result char(240);
rowseq number(8);

begin

rowseq:=1;
open pedigree_cursor;
fetch pedigree_cursor into pedigree_rec;
result:= 'pedigree0        Pedigree : ' ||
to_char(pedigree_rec.pedigree_id);
result := result || ': ' || pedigree_rec.pedigree_name;
insert into results(result,rowseq,process_id)
values (result,rowseq,&pid);
rowseq := rowseq+1;

/*
|| go get all the people
*/

open person_cursor;
loop
fetch person_cursor into person_rec;
exit when person_cursor %notfound;
result:= lpad(to_char(person_rec.person_id),5);
result:= result || lpad(to_char(person_rec.father_id),5);
```

Figure 3.13 Continued

```
result:= result || lpad(to_char(person_rec.mother_id),5);
result:= result || person_rec.sex;
result:= result || ' 1@@@@@@';
insert into results(result,rowseq,process_id)
values (result,rowseq,&pid);
rowseq := rowseq+1;
end loop;

/*
|| finished, so clean up
*/

close person_cursor;
close pedigree_cursor;
end;
/

/*
|| Generate the output file as an ordered sequence of lines
*/

select result from results
where process_id = &pid
order by rowseq
/

/*
|| Now clean up the results table
*/

delete from results
where process_id = &pid
/

/*
|| And make the changes permanent. This is faster than
|| a rollback.
*/

commit
/

quit

END
```

Figure 3.13   *SQL to generate input file for Pedigree/DRAW, from the data model developed in Figures 3.2 – 3.11.*

6   DISPLAYING AND MANAGING DATA

There are several good packages for displaying pedigrees. As for working with analytical programs, the principle is to transform data into the format required by these programs. As an example, code is provided for generating Pedigree/Draw input files (Figure 3.13).

For managing data, there is a wealth of application development environments. As an example, the Oracle SQL*Forms and SQL*Menu products were used to construct the LSHELL system (Bryant, 1994). Sybase and Informix provide their own development products, although it may be more appropriate, within a general laboratory or analyst setting, to consider using Open Data Base Connectivity (ODBC) as a glue to link client products that more easily integrate into a desktop environment. It is perfectly possible to link Microsoft Access under Windows 95 to Oracle Databases running under Unix, using ODBC. This means that data can easily be extracted and merged with other Windows applications, such as general statistical analysis programs or spreadsheets.

7   FURTHER DEVELOPMENT

The ideas developed in this treatment can easily be expanded to encompass the handling of quantitative traits and more complete integration of the analytical process. For ideas on how this might be achieved, refer to Bryant *et al.* (1997).

ACKNOWLEDGEMENTS

Many of the ideas presented in this chapter were created when the author was with the UK Imperial Cancer Research Fund, and the creative input of Anastassia Spiridou, supported by the European Commission (EC Grant GENE-C93-0003) is gratefully acknowledged. Gemini Research provided additional support during preparation of the manuscript.

REFERENCES

Bodmer WF, Bailey CJ, Bodmer J, Bussey HJR, Ellis A, Gorman P, Lucibello FC, Murday VA, Rider SH, Scambler P, Sheer D, Solomon E and Spurr NK (1987)

Localization of the gene for familial adenomatous polyposis on chromosome 5. *Nature* **328**, 614–616.

Bryant SP (1991) Software for Genetic Linkage Analysis. In *Methods in Molecular Biology 9: Protocols in Human Molecular Genetics*, pp. 403–418. (ed. C Mathew), Humana, New Jersey.

Bryant SP (1994) Genetic Linkage Analysis. In *Guide to Human Genome Computing*, pp. 59–110. (ed. MJ Bishop), Academic Press, London.

Bryant SP (1996) Software for Genetic Linkage Analysis – An Update. *Mol. Biotechnol.* **5**, 49–61.

Bryant SP, Spiridou A and Spurr NK (1997) The Integrated Genomic Database (IGD): enhancing the productivity of gene mapping projects. In *Proceedings of the 1996 International Symposium on Theoretical and Computational Genome Research*, pp. 117–134 (ed. S Suhai), Plenum, New York.

Buetow KH, Weber JL, Lundwigsen S, Scherpbier-Heddema T, Duyk GM, Sheffield VC, Wang Z and Murray JC (1994) Integrated human genome-wide maps constructed using the CEPH reference panel. *Nature Genet.* **6**, 391–393.

Chen WJ (1976) The Entity-Relationship Model – toward a unified view of data. *ACM Transactions on Database Systems* **1**, 9–36.

Curtis D (1990) A program to draw pedigrees using LINKAGE or LINKSYS data files. *Ann. Hum. Genet.* **54**, 365–367.

Erwin WG (1944) A pedigree of sex-linked recessive peroneal atrophy. *J. Hered.* **35**, 24–26.

Gyapay G, Morissette J, Vignal A, Dib C, Fizames C, Millasseau P, Marc S, Bernardi B, Lathrop M and Weissenbach J (1994) The 1993–94 Genethon human genetic linkage map. *Nature Genet.* **7**, 246–339.

Kelsell DP, Stevens HP, Ratnavel R, Bryant SP, Bishop DT, Leigh I and Spurr NK (1995) Genetic linkage studies in non-epidermolytic palmoplantar keratoderma: evidence for heterogeneity. *Hum. Mol. Genet.* **4**, 1021–1025.

Kruglyak L, Daly MJ, Reeve-Daly MP and Lander ES (1996) Parametric and nonparametric linkage analysis: a unified multipoint approach. *Am. J. Hum. Genet.* **58**, 1347–1363.

Lander ES and Green P (1987) Construction of multilocus genetic linkage maps in humans. *Proc. Natl. Acad. Sci. U.S.A.* **84**, 2363–2367.

Lathrop GM and Lalouel JM (1984) Easy calculations of lod scores and genetic risks on small computers. *Am. J. Hum. Genet.* **36**, 460–465.

Smith FJD, Early RAJ, MeMillan JR, Leigh IM, Geddes JF, Kelsell DP, Bryant SP, Spurr NK, Kirtschig G, Milana G, de Bono AG, Owaribe K, Wiche G, Pulkinnen L, Uitto J, Rugg EL, McLean WHI and Lane EB (1996) Plectin deficiency results in muscular dystrophy with epidermolysis bullosa. *Nature Genet.* **13**, 450–457.

Williams WR and Anderson DE (1984) Genetic epidemiology of breast cancer: segregation analysis of 200 Danish pedigrees. *Genet. Epidemiol.* **1**, 7–20.

4 Linkage Analysis Using Affected Sib-Pairs

Pak Sham and Jinghua Zhao

Department of Psychological Medicine, Institute of Psychiatry, De Crespigny Park, Denmark Hill, London SE5 8AF, UK

1 INTRODUCTION

In recent years, affected sib-pairs have become increasingly used for the linkage analysis of common diseases such as diabetes and asthma, with the purpose of localizing and characterizing the genetic components of these complex phenotypes. This increasing use of affected sib-pairs has been accompanied by the development of statistical methods and computer software for data analysis. This approach has already yielded a number of promising positive linkage findings, some of which are likely to be replicated and to lead to advances in our understanding of the aetiology of these disorders.

This chapter is an introductory account of non-parametric methods of linkage analysis for affected sib-pairs. First, the key concept of genetic identity-by-descent (IBD) will be explained. Next, some currently popular analytical methods and computer programs will be described. Finally, the strengths and limitations of the approach in general will be discussed. The reader is assumed to have an understanding of the principles of linkage analysis using the standard parametric (i.e. lod score) method.

2 GENETIC IDENTITY-BY-DESCENT BETWEEN SIB-PAIRS

Non-parametric linkage analysis is based on the detection of an association between the sharing of disease status and the sharing of marker alleles by relatives. There are two different measures of allele-sharing, identity-by-state (IBS) and IBD. Two alleles of the same form (i.e. having the same DNA sequence) are said to be IBS. If, in addition to being IBS, two alleles are descended from (and

Guide to Human Genome Computing, 2nd edition   Copyright © 1998 Academic Press Limited
ISBN 0-12-102051-7

are therefore replicates of) the same ancestral allele, they are said to be IBD. The general idea of non-parametric linkage analysis is that, in the vicinity of a disease locus, sib-pairs who are concordant for disease status (i.e. both affected or both unaffected) should show an increase in allele sharing, and those who are discordant for disease status (i.e. one affected and one unaffected) should show a decrease in allele sharing, from the level of allele sharing expected of sib-pairs. Of the two measures, IBS is easier to define and determine, but IBD is more directly relevant to linkage.

In the ideal situation, where the marker is extremely polymorphic so that all the alleles in the founding members of a pedigree are distinguishable from each other, IBS implies IBD, and the distinction between the two concepts becomes unnecessary. Consider two parents with genotypes S_1S_2 and S_3S_4. Mendel's law of segregation implies that each offspring of these two parents is equally likely to have the four genotypes S_1S_3, S_1S_4, S_2S_3, S_2S_4. The possible combinations of genotypes for a sib-pair, and the corresponding number of alleles IBS (and hence IBD), are as follows.

	Sibling 2			
Sibling 1	S_1S_3	S_1S_4	S_2S_3	S_2S_4
S_1S_3	2	1	1	0
S_1S_4	1	2	0	1
S_2S_3	1	0	2	1
S_2S_4	0	1	1	2

Each of these combinations has a probability of $(1/4)(1/4) = 1/16$, so that the probability of the sib-pair having 0, 1 and 2 alleles IBD are 4/16, 8/16 and 4/16 respectively. This 1/4, 1/2, 1/4 distribution for the IBD values of 0, 1, 2 represents the expected level of IBD for a sib-pair from which any excess or deficit in allele sharing must be measured.

In reality, DNA markers are never sufficiently polymorphic to ensure that all founder alleles are distinguishable from each other. This has the undesirable effect that the true IBD value (whether 0, 1 or 2) of a sib-pair cannot always be determined from the marker genotypes of the family. In this more realistic scenario, it becomes important to distinguish IBS from IBD, as the former no longer implies the latter. The distinction between IBS and IBD can be illustrated by the nuclear family shown in Figure 4.1.

The two offspring of this family are IBS for allele S_1. The parental origins of these two S_1 alleles are, however, different. Whereas the S_1 allele of the first offspring (genotype S_1S_2) is derived from the maternal S_1 allele, the S_1 allele of the second offspring (with genotype S_1S_3) is derived from the paternal S_1 allele. On the assumption that the two parents are indeed unrelated and non-inbred, these

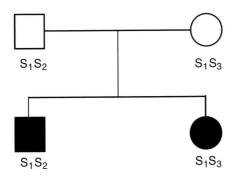

Figure 4.1 *Illustrative family.*

two S_1 alleles must be considered non-IBD. The sib-pair in this family is therefore IBS for 1 allele (S_1), but IBD for 0 allele.

To determine the IBS status of a sib-pair, it is sufficient to have genotypic data on the sib-pair alone. This is not true for IBD status. Suppose the parental genotypes for the illustrative family in Figure 4.1 are (S_1S_4, S_2S_3) instead of (S_1S_2, S_1S_3). In this case the two offspring must have inherited the same allele (S_1) from the father, so that the sib-pair is IBD for 1 allele. The genotypic data on the sib-pair (i.e. S_1S_2, S_1S_3) are therefore consistent with an IBD status of 0 or 1, depending on the genotypes of the parents. In general, if the IBS value of a sib-pair is 0, then the IBD value the sib-pair must also be 0, but if the IBS value of a sib-pair is 1 or 2, then the IBD value of the sib-pair cannot be determined without the parental genotypes.

3   GENETIC IDENTITY-BY-DESCENT AND RECOMBINATION FRACTION

Under the assumption that the two parents of a sib-pair are unrelated and non-inbred, the four parental alleles can be considered as four distinct ancestral alleles (whether or not they are IBS). The sib-pair can be IBD for 0, 1 or 2 alleles at each chromosomal location in the genome. The usefulness of IBD values for linkage analysis is based on the fact that the IBD values (0, 1 or 2) of a sib-pair are independent for non-syntenic loci, but are positively correlated for syntenic loci. The closer the two loci, the greater is the correlation between their IBD values. It can be shown that, when the recombination fraction between loci A and B is θ, the conditional distribution of IBD values at B (denoted as D_B), given the IBD value at A (denoted as D_A), is given in Table 4.1.

In this table Ψ is defined as $\theta^2 + (1 - \theta)^2$. From this it can be shown that the

Table 4.1 Conditional IBD distribution for sib-pair

	$P(D_B/D_A)$		
	$D_B = 0$	$D_B = 1$	$D_B = 2$
$D_A = 0$	Ψ^2	$2\Psi(1 - \Psi)$	$(1 - \Psi)^2$
$D_A = 1$	$\Psi(1 - \Psi)$	$1 - 2\Psi(1 - \Psi)$	$\Psi(1 - \Psi)$
$D_A = 2$	$(1 - \Psi)^2$	$2\Psi(1 - \Psi)$	Ψ^2

correlation between D_A and D_B is $2\Psi - 1$. This correlation ranges from the value of 0 when $\theta = 1/2$, to the value of 1 when $\theta = 0$. The positive correlation between the IBD values of two tightly linked loci has the important implication that an affected sib-pair is expected to show an increase in IBD not only at the disease locus itself but also at neighbouring markers.

4   IDENTITY-BY-DESCENT DISTRIBUTION FOR AN AFFECTED SIB-PAIR

The most popular family unit for non-parametric linkage analysis is an affected sib-pair, with or without parents. Affected sib-pairs are often far more informative for linkage than unaffected sib-pairs, or sib-pairs with one affected and one unaffected member, under many plausible models of complex inheritance. Intuitively, this is because the presence of disease is often more indicative of the presence of the disease gene than the absence of the disease is indicative of the absence of the disease gene. Moreover, if the disease gene is rare, it is absent in most parents whose offspring are all unaffected. Most families with two unaffected offspring are therefore uninformative for linkage.

A more precise analysis of the IBD distribution of an affected sib-pair is desirable for two purposes. The first is that the IBD distribution may provide insights that are helpful in the construction of statistical tests for linkage. The second is that it can be used to calculate expected IBD values that are necessary for power analysis and sample size determination. The greater the deviation of the expected IBD distribution from the null hypothesis values of 1/4, 1/2, 1/4, the greater is the power for detecting linkage.

The IBD distribution of an affected sib-pair at a marker locus linked to a disease locus was derived by Suarez *et al.* (1978) under the assumptions of a single-locus disease model and random mating. The parameters of the model are the disease allele frequency (p), the penetrances (f_0, f_1 and f_2, defined as the conditional probabilities of the disease given 0, 1 and 2 copies of the disease allele), and the recombination fraction (θ). The following quantities are then defined in terms of these parameters:

$$q = 1 - p$$
$$K = p^2 f_2 + 2pq f_1 + q^2 f_0$$
$$V_A = 2pq[p(f_2 - f_1) + q(f_1 - f_0)]^2$$
$$V_D = p^2 q^2 (f_2 - 2f_1 + f_0)^2$$
$$\Psi = \theta^2 + (1 - \theta)^2$$

where K is the population prevalence of the disease, and V_A and V_D are additive and dominance variances respectively. Suarez *et al.* (1978) showed that the IBD probabilities for an affected sib-pair are related to these quantities as follows:

$$P(\text{IBD} = 0) = \frac{1}{4} - \frac{(\Psi - 0.5)V_A + (2\Psi - \Psi^2 - 0.75)V_D}{4(K^2 + 0.5 V_A + 0.25 V_D)}$$

$$P(\text{IBD} = 1) = \frac{1}{2} - \frac{2(\Psi^2 - \Psi + 0.25)V_D}{4(K^2 + 0.5 V_A + 0.25 V_D)}$$

$$P(\text{IBD} = 2) = \frac{1}{4} + \frac{(\Psi - 0.5)V_A + (\Psi^2 - 0.25)V_D}{4(K^2 + 0.5 V_A + 0.25 V_D)}$$

Holmans (1993) and Faraway (1993) showed that only limited combinations of IBD probabilities are compatible with these equations. If the IBD probabilities are

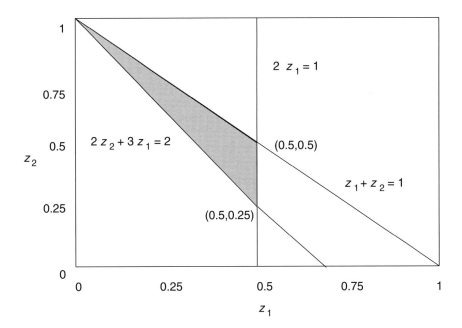

Figure 4.2 *The possible triangle for IBD probabilities.*

denoted as z_0, z_1 and z_2, the combinations of IBD values within the triangular region defined by $z_1 \leqslant 1/2$, $(z_1 + z_2) \leqslant 1$ and $(3z_1/2 + z_2) \geqslant 1$ (Figure 4.2) are compatible, whereas those outside the triangle are not compatible, with a single locus disease model. If it is assumed further that V_D (dominance variance) is zero, then the only admissible combination of IBD values are on the line $z_1 = 1/2$, for $1/4 \leqslant z_2 \leqslant 1/2$.

The IBD distribution at the disease locus (i.e. $\theta = 0$) can be re-expressed in terms of the recurrence risks of the disease in various classes of relatives (e.g. monozygotic twins, siblings, offspring) of affected probands (Risch, 1990a). If the recurrence risk of the disease in a relative of class R is denoted as K_R, and the ratio of this risk to the population risk (K) is denoted as λ_R, where the subscript R may take the values M (monozygotic twin), S (sibling), O (offspring), then the IBD probabilities of an affected sib-pair can be rewritten in terms of the 'relative risks' λ_M, λ_S, λ_O as follows:

$$P(\text{IBD} = 0) = \frac{1}{4}\,\frac{1}{\lambda_S}$$

$$P(\text{IBD} = 1) = \frac{1}{2}\,\frac{\lambda_O}{\lambda_S}$$

$$P(\text{IBD} = 2) = \frac{1}{4}\,\frac{\lambda_M}{\lambda_S}$$

As these probabilities must sum to 1, it follows that (under a single locus disease model and random mating):

$$\lambda_M = 4\lambda_S - 2\lambda_O - 1$$

The extension of this formulation to multilocus disease models is simple, especially in cases where the effects of the alleles on risk are multiplicative. In this case, each locus contributes a 'factor' to the relative risk of each class of relative:

$$\lambda_R = \lambda_{1R}\lambda_{2R}\lambda_{3R}\cdots$$

The IBD distribution at the ith disease locus is then:

$$P(\text{IBD}_i = 0) = \frac{1}{4}\,\frac{1}{\lambda_{iS}}$$

$$P(\text{IBD}_i = 1) = \frac{1}{2}\,\frac{\lambda_{iO}}{\lambda_{iS}}$$

$$P(\text{IBD}_i = 2) = \frac{1}{4}\,\frac{\lambda_{iM}}{\lambda_{iS}}$$

with the implication that

$$\lambda_{iM} = 4\lambda_{iS} - 2\lambda_{iO} - 1$$

for each constituent locus. The calculation of the IBD distribution at a particular disease locus requires certain assumptions to be made about the 'relative risks' (i.e. λ_{iM}, λ_{iS}, λ_{iO}) associated with that locus.

Example. Suppose that the empirical relative risks of a disease in monozygotic cotwins, siblings and offspring of affected probands are 45, 10 and 10, and that the 'relative risks' of all contributory loci are identical. Under these assumptions, the number of loci, k, must satisfy the relationship $(45)^{1/k} = 4(10)^{1/k} - 2(10)^{1/k} - 1$. The solution of this equation yields $k = 4.4$, when λ_{iM}, λ_{iS} and λ_{iO} are 2.375, 1.688 and 1.688 for each constituent locus. Under these 'relative risks', affected sib-pairs can be shown to have a probability distribution of $z_0 = 0.148$, $z_1 = 0.5$ and $z_2 = 0.352$ for sharing 0, 1 and 2 alleles IBD, at each constituent locus. If a simple test of proportion of $z_2 = 0.25$ against $z_2 > 0.25$ is adopted, then the expected value of z_2 (0.352) would indicate that a sample size of 420 affected sib-pairs (for whom IBD status can be determined) is required for 80% power to detect linkage at a significance level of 0.0001.

5   METHODS OF AFFECTED SIB-PAIR ANALYSIS

There are several variations of the affected sib-pair method, which can be classified broadly into those based on (or closely related to) Pearson χ^2 statistics (counting the numbers of sib-pairs sharing 0, 1, and 2 alleles IBD and comparing these numbers to their expectations) and those based on likelihood ratio statistics (treating IBD probabilities as unknown parameters).

5.1   Simple IBD Counting Methods

In early versions of the counting methods, families are retained only if the parental genotypes always allow the IBD status of the sib-pair to be determined fully. The necessary conditions for this are that both parents must be heterozygous and that the two parents must not have the same genotype. Let the total number of affected sib-pairs in such families be n, and numbers of pairs with IBD values of 0, 1 and 2 be n_0, n_1 and n_2 ($n = n_0 + n_1 + n_2$). Several simple test statistics have been proposed to assess whether these numbers deviate from the expected proportions of (1/4, 1/2, 1/4). The simplest of these is the ordinary Pearson χ^2 statistic:

$$S_1 = \Sigma_i \frac{(n_i - e_i)^2}{e_i}$$

where e_0, e_1 and e_2 are $n/4$, $n/2$ and $n/4$ respectively. This statistic can be referred to a χ^2 distribution with two degrees of freedom for a test of linkage.

Another family of test statistics are linear combinations of the form $vn_1 + n_2$ (Knapp *et al.*, 1994). Among these statistics, the most popular is the so-called 'mean test', which is obtained by setting $v = 1/2$. As the mean and variance of $(1/2)n_1 + n_2$ under the null hypothesis are $n/2$ and $n/8$ respectively, the test statistic

$$S_2 = \frac{(\frac{1}{2}n_1 + n_2) - \frac{n}{2}}{\sqrt{\frac{n}{8}}}$$

has asymptotically a standard normal distribution under the null hypothesis of non-linkage. This is simply a test of whether the proportion of alleles IBD is $1/2$. As the alternative hypothesis of linkage implies $S_2 > 0$, the test is one-tailed. This test can be shown to be locally most powerful (i.e. for θ close to $1/2$) under all possible single-locus models of inheritance, and to be uniformly most powerful for all value of θ when $f_1^2 = f_0 f_2$ (Knapp *et al.*, 1994). In simulation studies, the mean test appears to perform adequately under a wide range of conditions (Suarez and Van Eerdewegh, 1984; Blackwelder and Elston, 1985).

5.2   Partially Informative Families

The restriction of affected sib-pair analysis to 'fully informative' families can result in some loss of information. In some families, for example, those with parental genotypes $A_1A_2 \times A_3A_3$, IBD status can be deduced for the alleles transmitted from the A_1A_2 parent but not the A_3A_3 parent. In other families, the genotype of an individual (for example, a parent) may be unavailable, but the available genotypes in other family members may allow the probability distribution of the missing genotype to be calculated. An extension of the classical affected sib-pairs method to make use of such partial information was developed by Sandkuijl (1989), and implemented in the ESPA (Extended Sib-Pair Analysis) program. Another program that uses a similar approach is SIBPAIR in the ANALYSE package written by Terwilliger.

For each affected sib-pair, the ESPA program attempts a probabilistic reconstruction of missing parental or sibling genotypes from the known genotypes of the family members. In practice, this involves calculating the probabilities of the possible genotype configurations, given the known genotypes in the pedigree, using the MLINK program. For each possible configuration, the number of alleles IBD, the number of alleles non-IBD, and the number of alleles whose IBD status is unknown, are counted. We denote these three counts, for configuration i, by n_{i1}, n_{i0} and $(2 - n_{i1} - n_{i0})$. The 'estimated' counts of alleles IBD and alleles non-IBD in the sib-pair are then the weighted sums $\Sigma_i p_i n_{i1}$ and $\Sigma_i p_i n_{i0}$, where p_i is the probability of configuration i. These counts are then aggregated across all the affected sib-pairs in the sample, to obtain the total 'estimated' counts of alleles IBD and alleles non-IBD. If these counts are denoted as N_1 and N_0, then ESPA computes the statistic $(N_1 - N_0)^2/(N_1 + N_0)$ and interprets it as a one-tailed χ^2 test for the hypothesis of linkage.

The ESPA program has many attractive features. It uses standard LINKAGE format pedigree and parameter input files. The final counts N_1 and N_0, and the test statistic $(N_1 - N_0)^2/(N_1 + N_0)$ are easy to interpret, and the program handles pedigree data with multiple affected sib-pairs and missing genotypes. However, it is unclear whether the statistic $(N_1 - N_0)^2/(N_1 + N_0)$ is asymptotically χ^2 as the counts N_1 and N_0 are 'estimates' rather than observations. Moreover, multiple affected sib-pairs in the same family, which may not constitute independent observations, are included in the calculation of N_1 and N_0. This non-independence may distort the distribution of the test statistic. Also, the reconstruction of missing parental genotypes from the observed genotypes of the affected sib-pair may cause the test to be biased in favour of the null hypothesis. This is because the reconstruction of parental genotypes is easier for sib-pairs who are non-IBD for both alleles, than for sib-pairs who are IBD for both alleles. Thus, an affected sib-pair with four distinct alleles (e.g. A_1A_2, A_3A_4) will 'force' the two parents to have these four alleles even if parental genotypes are unavailable, so that the sib-pair will be given full weight in the analysis. On the other hand, an affected sib-pair who are both homozygous with the same allele (e.g. A_1A_1, A_1A_1) may be IBD for both alleles, but whether this is so will be uncertain unless the parental genotypes are available. The weights given to such sib-pairs will therefore be incomplete. The net effect of this bias is a spurious excess of non-IBD alleles. This problem can be avoided by ignoring the genotypes of the affected sib-pair when reconstructing the genotypes of the parents. However, this may result in some loss of information, which may be even more detrimental than the bias in favour of the null hypothesis, especially if the marker is highly polymorphic.

5.3   Likelihood Methods

Likelihood methods of affected sib-pair analysis were introduced to provide a more satisfactory way of dealing with the problem of incomplete data (Risch, 1990b; Holmans, 1993). The statistical model is that each affected sib-pair has probabilities z_0, z_1 and z_2 of having 0, 1 and 2 alleles IBD. These probabilities, written as the vector z, are defined as the parameters of the model. The likelihood of a set of genotypic data x can be written as a function of z (and the allele frequencies at the marker locus, which are either known or estimated from the data). For a family containing one affected sib-pair, the likelihood can be decomposed as follows:

$$L(z) = \Sigma_{i=0}^2 \; z_i P(x|\text{IBD} = i)$$

which is a linear combination of z_0, z_1 and z_2. If the log-likelihood maximized with respect to z is denoted as $\ln L_1$ and the log-likelihood at the null hypothesis ($z_0 = 0.25$, $z_1 = 0.5$, $z_2 = 0.25$) is denoted as $\ln L_0$, then the likelihood ratio statistic $2(\ln L_1 - \ln L_0)$ is asymptotically χ^2 with 2 degrees of freedom, and provides a test for linkage. There are only two free parameters because of the obvious constraint $z_0 + z_1 + z_2 = 1$.

Two levels of restriction are often imposed on z. Holmans (1993) suggested constraining the parameters z_1 and z_2 to the triangular region compatible with a single locus disease model (Figure 4.2). Let the maximum log-likelihood within this region be $\ln L_1^*$, then a likelihood ratio test statistic for linkage is $2(\ln L_1^* - \ln L_0)$. Using the method of Self and Liang (1987), the asymptotic distribution of this statistic was shown to be a mixture of χ^2 distributions with 0, 1 and 2 degrees of freedom (a χ^2 with 0 degrees of freedom is defined as 0). The mixing proportions depend on marker allele frequencies, but are approximately 0.41, 0.50, 0.09.

The second level of restriction on z is obtained by the assumption of zero dominance variance, which constrains z_1 to be $1/2$, and z_2 to between $1/4$ and $1/2$. Let the maximum log-likelihood on this line be $\ln L_1^{**}$, then a likelihood ratio test statistic for linkage is $2(\ln L_1^{**} - \ln L_0)$. The asymptotic distribution of this statistic is a $50:50$ mixture of χ^2 distributions with 0 and 1 degrees of freedom.

Programs that implement the likelihood method include SPLINK and MAP-MAKER/SIBS, with the latter being able to deal with multipoint data (Kruglyak and Lander, 1995). In multipoint analysis, the likelihood function at a chromosomal location is the probability of the genotypic data of the entire marker set in the region over the IBD values at that chromosomal location. The ability of MAPMAKER/SIBS to handle multipoint data enables full IBD information to be extracted from a region, making it the program of choice.

MAPMAKER/SIBS uses input pedigree and parameter files in standard LINKAGE format, and can be set to either of the two levels of restriction on z. It can be used to compute a series of likelihood ratio statistics across a chromosomal region, in order to detect areas with evidence of excess allele sharing. It can also be used to perform exclusion mapping given a set of hypothetical values for z.

6   AFFECTED SIB-PAIR ANALYSIS: STRENGTHS AND LIMITATIONS

The linkage information content of affected sib-pairs, compared with other types of affected relative pairs, and with pedigrees with a greater number of affected members, is an important issue for the design of linkage studies. Generally speaking, for a gene of major effect (i.e. high penetrance, high 'relative risk'), affected sib-pairs contain far less linkage information than more distantly related affected relative pairs or large pedigrees with multiple affected members. As the 'relative risk' decreases, however, affected sib-pairs become progressively more attractive, as aetiological heterogeneity become more likely for distantly related affected relative pairs, and parental homozygosity at the disease locus becomes more likely in densely affected pedigrees. Risch (1990a) showed that, at a 'sibling relative risk' of 2 or less, affected sib-pairs are equally informative for linkage as other more distantly related affected relative pairs.

The increasing recognition that common genes of minor or modest effects may be involved in common diseases such as diabetes and depression has led to the increasing use of the affected sib-pair approach. However, it does not appear sensible to adhere strictly to affected sib-pairs, to the exclusion of other potentially informative family units. In other words, it would be wasteful not to include other types of affected relative pairs, or families with three or more affected members, if these were encountered in the process of identifying affected sib-pairs. As the affected sib-pair methods described above are designed to deal with pedigrees with just two affected siblings, the inclusion of other family types presents new analytical problems that are still currently under investigation. One possible approach is the generalization of statistics based on excess allele sharing to general pedigrees (Curtis and Sham, 1994; Davis *et al.*, 1996; Kruglyak *et al.*, 1996). Another possible approach is to modify the lod score method by treating allele frequencies and penetrances as nuisance parameters (Curtis and Sham, 1995). At present, the most popular test is probably that based on the NPL_{all} statistic of the GENEHUNTER program. This statistic is a standardized measure of excessive IBD allele sharing among the affected members in pedigrees.

Another limitation of the affected sib-pair method is that it does not make full use of information from unaffected family members. When the penetrance of the disease allele is low, many unaffected individuals will be non-penetrant carriers of the disease allele, so that the loss of linkage information is expected to be small. If, however, there is a quantitative trait which is affected by the presence or absence of the disease allele, then the value of the trait may provide information about the disease genotype in unaffected individuals. The analysis of sib-pair data for linkage between a quantitative trait and marker loci is known as QTL (quantitative trait loci) linkage analysis (Fulker *et al.*, 1995; Risch and Zhang, 1995).

Finally, it must be recognized that linkage analysis is inherently limited by the effect size of the disease gene. When the disease allele increases the risk of the disease by a factor less than 2, then the detection of linkage will require enormous samples, whatever method of linkage analysis is used. The same effect may, however, be detectable by allelic association analysis in a realistic sample (Risch and Merigangas, 1996).

7   CONCLUSION

The affected sib-pair method of linkage analysis is useful primarily for common diseases where the 'sibling relative risk' associated with any contributory locus is modest $(1.5 < \lambda_{iS} < 2.5)$. Below a 'sibling relative risk' of 1.5, the number of affected sib-pairs required to detect linkage becomes prohibitively large. On the other hand, when the 'sibling relative risk' is above 2.5, other types of affected relative-pairs and larger pedigrees become substantially more informative for linkage than affected sib-pairs.

For a sample that consists predominantly of affected sib-pairs (with or without unaffected relatives), the analytical method and computer program of choice is probably MAPMAKER/SIBS. This method is able to make full use of multipoint and incomplete data. If the sample contains an appreciable proportion of families with more complex structure, then an alternative method such as the NPL_{all} test in GENEHUNTER is advisable.

REFERENCES

Blackwelder WC and Elston RC (1985) A comparison of sib-pair linkage tests for disease susceptibility loci. *Genet. Epidemiol.* **2**, 85–97.

Curtis D and Sham PC (1994) Using risk calculation to implement an extended relative pair analysis. *Ann. Hum. Genet.* **58**, 151–162.

Curtis D and Sham PC (1995) Model-free linkage analysis using likelihoods. *Am. J. Hum. Genet.* **57**, 703–716.

Davis S, Schroeder M, Goldin LR and Weeks DE (1996) Nonparametric simulation-based statistics for detecting linkage in general pedigrees. *Am. J. Hum. Genet.* **58**, 867–880.

Faraway JJ (1993) Improved sib-pair linkage test for disease susceptibility loci. *Genet. Epidemiol.* **10**, 225–233.

Fulker DW, Cherny SS and Cardon LR (1995) Multipoint interval mapping of quantitative trait loci, using sib-pairs. *Am. J. Hum. Genet.* **56**, 1224–1233.

Holmans P (1993) Asymptotic properties of affected-sib-pair linkage analysis. *Am. J. Hum. Genet.* **52**, 362–374.

Knapp M, Seuchter SA and Baur MP (1994) Linkage analysis in nuclear families. 1. Optimality criteria for affected sib-pair tests. *Hum. Hered.* **44**, 37–43.

Kruglyak L and Lander ES (1995) Complete multipoint sib-pair analysis of qualitative and quantitative traits. *Am. J. Hum. Genet.* **57**, 439–454.

Kruglyak L, Daly MJ, Reeve-Daly MP and Lander ES (1996) Parametric and non-parametric linkage analysis: a unified multipoint approach. *Am. J. Hum. Genet.* **58**, 1347–1363.

Risch N (1990a) Linkage strategies for genetically complex traits. II. The power of affected relative pairs. *Am. J. Hum. Genet.* **46**, 229–241.

Risch N (1990b) Linkage strategies for genetically complex traits. III. The effect of marker polymorphism on analysis of affected relative pairs. *Am. J. Hum. Genet.* **46**, 242–253.

Risch N and Merigangas K (1996) The future of genetic studies of complex human diseases. *Science* **273**, 1516–1517.

Risch N and Zhang H (1995) Extreme discordant sib-pairs for mapping quantitative trait loci in humans. *Science* **268**, 1584–1589.

Sandkuijl LA (1989) Analysis of affected sib-pairs using information from extended families. Multipoint mapping and linkage based upon affected pedigree members: *Genetic Analysis Workshop 6*, Alan R Liss, New York.

Self SG and Liang KY (1987) Asymptotic properties of maximum likelihood estimators and likelihood ratio tests under nonstandard conditions. *J. Am. Stat. Assoc.* **82**, 605–610.

Suarez BK and Van Eerdewegh P (1984) A comparison of three affected-sib-pair scoring methods to detect HLA-linked disease susceptibility genes. *Am. J. Med. Genet.* **18**, 135–146.

Suarez BK, Rice JP and Reich T (1978) The generalised sib-pair IBD distribution: its use in the detection of linkage. *Ann. Hum. Genet.* **42**, 87–94.

LIST OF COMPUTER PROGRAMS

ANALYSE
Author: Joe Terwilliger
ftp://ftp.well.ox.ac.uk/pub/genetics/analyse or
ftp://linkage.cpmc.columbia.edu/software/analyse

APM
Authors: Daniel E. Weeks (University of Oxford and University of Pittsburg, dweeks@watson.hgen.pitt.edu), Mark Schroeder (mark@holmes.hgen.pitt.edu)
ftp://watson.hgen.pitt.edu/pub/apm

ASPEX
Authors: David Hinds (dhinds@lahmed.stanford.edu), Neil Risch (Stanford University),
http://lahmed.stanford.edu/pub/aspex/index.html
ftp://lahmed.stanford.edu/pub/aspex (the utility scripts are written in Perl)
(the graphic script requires xmgr, available from
ftp://ftp.teleport.com/pub/users/pturner/acegr)

ERPA
Author: Dave Curtis (dcurtis@hgmp.mrc.ac.uk)
http://www.gene.ucl.ac.uk/~dcurtis/software.html
ftp://ftp.gene.ucl.ac.uk/pub/packages/dcurtis

ESPA
Author: Lodewijk Sandkuijl (sandkuijl@rullf2.leidenuniv.nl)

GAS

Author: Alan Young (ayoung@vax.ox.ac.uk)
http://users.ox.ac.uk/~ayoung/gas.html, and a US mirror site:
http://linkage.rockefeller.edu/soft/gas/gas2.html
ftp://ftp.ox.ac.uk/pub/users/ayoung, and a US mirror site:
ftp://linkage.rockefeller.edu/software/gas

MAPMAKER/SIBS

Authors: Leonid Kruglyak (leonid@genome.wi.mit.edu), Mark Daly
(mjdaly@genome.wi.mit.edu), Eric Lander (lander@genome.wi.mit.edu)
(Whitehead Institute)
ftp://ftp-genome.wi.mit.edu/distribution/software/sibs

SAGE

Authors: R. Elston, J. Bailey-Wilson, G. Bonney, L. Tran, B. Keats, A. Wilson
http://darwin.mhmc.cwru.edu/ (previously: http://csmith.biogen.lsumc.edu)
ftp: [not in public domain]

SIMIBD

Authors: Sean Davis (davis@moriarty.hgen.pitt.edu), Daniel E. Weeks
ftp://watson.hgen.pitt.edu/pub/simibd

SPLINK

Author: David Clayton, MRC Biostatistics Unit, Cambridge
ftp.mrc-bsu.cam.ac.uk (under/pub/methodology/genetics)

5 Comparative Mapping in Humans and Vertebrates

Martin J. Bishop

Medical Research Council, UK Human Genome Mapping Project Resource Centre, Hinxton, Cambridge CB10 1SB, UK

1 CONSERVATION OF GENES

1.1 Genes Conserved Across Diverse Phyla

It is believed that vertebrates share much the same set of about 100 000 genes, although there will be genes special to each of the vertebrate classes. Indeed, similarities of human genes with fly (*Drosophila*), nematode (*Caenorhabditis*), yeast (*Saccharomyces*) and even bacterial (*Escherichia*) genes are shedding light on gene function and human disease. A database cross-referencing the genetics of model organisms with mammalian phenotypes is available from the National Center for Biotechnology Information (NCBI) (Bassett *et al.*, 1995). It may be accessed at http://www.ncbi.nlm.nih.gov/XREFdb/. At April 1997 over 80 human diseases had been related to mutated proteins with homologues in the model organisms.

Approximately two-thirds of the genes discovered in the yeast and nematode sequencing projects are new and unrelated to each other or to human expressed sequence tags (ESTs). This suggests that the majority of genes with recognizable similarities across phyla are already known (Green *et al.*, 1993). The total number of ancient conserved regions in proteins is estimated to be 860. The conclusion is that two-thirds of genes are either phylum specific or are evolving so rapidly that distant relatives are no longer recognizable on the basis of sequence.

1.2 Conservation of Vertebrate Genes

There are three major groups of living mammals: the Eutheria, Metatheria and

Guide to Human Genome Computing, 2nd edition
ISBN 0-12-102051-7

Prototheria. The orders of the Eutheria diverged within a short period 60–80 million years ago and the four orders discussed in Section 4 are equally distant (as far as is known) (Miyamoto, 1996). The eutherian placental mammals diverged from the marsupial metatherian mammals 150 million years ago and their common ancestor diverged from the egg-laying prototherian mammals 170 million years ago. According to fossil evidence, mammals diverged about 200 million years ago from a branch of the reptiles that has left no other descendants. Mammals, birds and the major reptilian groups are equally distant and share a common ancestor 350 million years ago. Reptiles diverged from fish about 400 million years ago. It is believed that at least two rounds of tetraploidization occurred during the evolution of fish from primitive chordates (Lundin, 1993). This view is supported by the numbers of clusters of *Hox* genes, one in amphioxus, two in lamprey and four in vertebrates. This implies that it is meaningful to study conservation of duplicated linkage groups within species as well as between species. Gene loss after duplication is a key feature of both tetrapod and fish evolution (Aparicio *et al.*, 1997).

When an attempt is made to examine molecular data for evidence of the course of vertebrate evolution it is apparent that there is very little information at present. For example, starting from amphioxus proteins and attempting to find proteins to compare in the vertebrate classes yielded a single dataset (Figure 5.1) resulting from the work of Smith and Doolittle (1992). These proteins are the Mn/Fe superoxide dismutases that place primates (human) with rodents and show the mammalian orders (rodents, artiodactyls and lagomorphs) diverging as much from each other as from lamprey, hagfish and amphioxus. Obviously, such fragmentary information does not contain an accurate record of evolution, but the protochordate and chordates do group together compared with a bacterium, a yeast, a plant and two invertebrates.

1.3 Estimation of Gene Identity Between Species

Many criteria may be used to characterize genes or parts of genes, but the two most fundamental are DNA sequence and order in relation to other genes on the chromosome. Some genes occur in many copies in a single organism (e.g. rRNA genes). The majority of genes are not single copy but exist in families that are hypothesized to have originated from an ancestral gene by duplication and subsequent divergence. Duplications may be tandem duplication of local regions, or may be more extensive involving whole chromosomes or even whole genomes. Subsequent deletions can give rise to gene families of great complexity, with members missing in some taxa. Related genes may alter each other by a process known as gene conversion.

Except in the case of the sex chromosomes, a mammalian individual has two copies of each gene, one from its mother and one from its father. In comparative

```
Man          HINAQIMQLH HSKHHAAYVN NLNVTEEKYQ EAL------- AKGDVTAQIA LQPALKFNGG
Guinea pig   ....E..... ........L. ...IA..... .......... ........V. ..........
Mouse        .......... .......... ...A.....H .......... .....T.V. ..........
Rat          .......... .....T... .........H .......... ......T.V. ..........
Rabbit       .......E.. .......... ...A.....R .......... .R.....HV. .......K..
Ox           .......... .......... ...A.....R .......... E......... ..........
Pig          .......... ..E....... .......V.. .......... K......V. ..........
Lamprey      ..S.N.... ......T... ....A.Q.LA ..V....... .......E.. ....I.....
Hag fish     ..S.E.... .......... .V..V...LA .......... G....NT.VS ....FR....
Amphioxus    V.S.E...V. .Q....T... ...AA..QLA ..I....... H.Q...KM.. ..S.I.....
Fruit fly    I.CRE..E.. .Q...QT... ...AA..QLE ..K....... S.S.T.KL.Q .A...R....
Nematode     V.SHE..... .Q....T... ...QI...LH ..V....... S..N.KEA. ..........
Maize        A.SGE..R.. .Q....T..A .Y.KAL.QLE T.V....... S...AS.VVQ ..A.I.....
Yeast        Y.SG..NE.. YT...QT... GF.TAVDQF. .LSDLLAKEP SPANARKM.. I.QNI..H..
E. coli      .FDK.T.EI. .T...QT... .A.AAL.SLP .FANLPVEEL ITKLDQLPAD KKTV.RN.A.

Man          GHINHSIFWT NLSPNGGG-- -EPKGELLEA IKRDFGSFDK FKEKLTAASV GVQGSGWGWL
Guinea pig   .......... .......... .......... .......... .......V.. ..........
Mouse        .....T.... ....K..... .......... .......... E. ........M.. ..........
Rat          .......... ....K..... .......... .......... E. .......V.. ..........
Rabbit       .....T.... .......... .......... .......... ...R...V.. ..........
Ox           .......... .......... ...Q...... .......... A. .......V.. ..........
Pig          .......... .......... .......... .......... E. .......V.. ..........
Lamprey      .......... .......... .A.T.D.QK. .ET....T. LQ..MS.V.. A.........
Hag fish     .........R ....S..... .Q.C.D..K. .EN....V.. LR...V..A. .......A..
Amphioxus    .........N ..C.S..... ...T.P.A.. .T.....EA ....M...T. A.........
Fruit fly    .....T...Q .....KT-.. .Q.SDD.KK. .ESQWK.LEE .KE..TLT. A.........
Nematode     .......... ..AKD..-.. ...SA...T. ..S....L.N LQKQ.S.ST. A.........
Maize        ..V......K .K.ISE.GG EP.H.K.GW. .DE.....EA LVK.MN.EGA AL.....V..
Yeast        .FT..CL..E ..A.ESQ.GG EP.T.A.AK. .DEQ...L.E LIKLTNTKLA .......AFI
E. coli      ..A...L..K G.KKGTT-.. .-LQ.D.KA. .E.....V.N ..AEFEK.AA SRF....A..

Man          GFNKERG-HL QIAACPNQD- PLQGT-TG-- -LIPLLGIDV -
Guinea pig   ........C. .....S.... .......... .......... .
Mouse        .....Q..R. .....S.... .......... .......... .
Rat          .....Q..R. .....S.... .......... .......... .
Rabbit       .....Q.... .....A.... .......... .......... .
Ox           .....Q..R. .....S.... .......... .......... .
Pig          .....Q..R. .....S.... .......... ..V....... .
Lamprey      .YD..T..R. R....A.... ...A...... .......... .
Hag fish     .....SK.R. ...T.A.... .......... ..F....... .
Amphioxus    .LDPTSK.K. R.V....... ..E....... ..K....... .
Fruit fly    ...KS..K. .L..L..... ..EAS..... .....F.... .
Nematode     .YCPKGK.I. KV.T.A.... ..EA...... ..V..F.... .
Maize        ALD..AK.KV SVETTA.... ..VTKGAS.. ..V....... .
Yeast        VK.LSN.GK. DVVQTY.... TVT.P.--.. ..V..VA..A W
E. coli      VLK---.DK. AVVSTA...S ..M.EAISGA SGF.IM.L.. .
```

Figure 5.1   *Partial sequences of some Mn/Fe superoxide dismutase proteins.*

mapping discussions a typological approach is usually adopted. That is, there exists one identifiable gene sequence per gene per species. In reality, genes show considerable within-species variation as we know from examples of medical importance (e.g. haemoglobinopathies, cystic fibrosis). Much greater differences exist between aberrant human haemoglobins than between human and chimp 'type' sequences. For complete understanding, a knowledge of within-species variation is necessary.

Given two gene sequences, we can measure their similarity. There are an infinite number of measures of gene similarity and we need to choose one which is of practical utility for comparative mapping purposes. The measure of similarity need have no evolutionary model associated with it. It is possible to recognize proteins with distantly similar sequences or very similar structures that are not related by descent from a common ancestral protein (hypothesis of convergent evolution).

When similar sequences are believed to be related to each other by evolutionary descent, they are said to be homologous. Paralogous genes is the term used for genes possessed by an individual that are related by duplication and divergence. Pseudogenes without introns may result from incorporation of processed RNA messages back into genomic DNA by reverse transcription. Shuffling and reuse of exons appear to occur. The whole picture is, therefore, one of considerable complexity. Orthologous genes is the term used for genes shared by two individuals which are related by divergence following reproductive isolation.

Homologies, orthologies and paralogies are not facts but evolutionary inferences under a model of evolution in the light of the data observed today. A bifurcating tree of gene duplications can represent the relationships and divergence times between paralogous genes in an individual. Similarly, a bifurcating tree of reproductive isolation events can represent the relationships of orthologous genes for a number of species. Unique orders of gene duplication events in a lineage and of speciation events exist historically but are unknown today. Their inference from the sequence data is complex and difficult. Morphological, including fossil, evidence is used to constrain the solutions. Horizontal transmission of genetic material, for example by infectious elements, will confuse the situation. In the absence of complete information about the members of a gene family from the individuals being compared, further confusion can readily arise.

The Workshop on Comparative Genome Organization (1996) defined working criteria for the establishment of homology between genes mapped in different species (Table 5.1). Gene nomenclature undergoes a process of refinement as knowledge advances. A comparative element is involved in identifying the genes for producing single-species gene maps.

2 CONSERVATION OF LINKAGE GROUPS

2.1 Determination of Gene Order in a Single Species

Gene order in a single species may be determined by a variety of methods, often complementary, many of which are the subjects of other chapters in this book. Genetic linkage mapping requires polymorphic markers and a suitable cross (animals) or families (humans). It has provided reliable framework maps for a number of mammalian species. Radiation hybrid mapping (Chapter 6) does not

Table 5.1 Criteria for establishing homology between genes mapped in different species (Comparative Genome Organisation, 1996)

	Stringent	Weak
Gene	Similar nucleotide sequence Cross-hybridization to probe Conserved map position	Similar transcription profile
Protein	Similar amino acid sequence	Similar subunit structure Immunological cross-reaction Similar expression profile Similar subcellular localization Similar substrate specificity Similar response to inhibitors
Phenotype	Complementation of function	Similar mutant phenotype

require polymorphic markers and can more easily provide good resolution. Fluorescent *in situ* hybridization (FISH) is a cytogenetic technique that can give visual confirmation of the established framework map for species in which the individual chromosomes can readily be distinguished. The ultimate dataset is the DNA sequence for each chromosome produced from the sequence ready clone map (Chapter 7). Gene identification then has to be undertaken (Chapter 10) and is considerably aided by inter-specific comparisons.

This idealized scenario of complete sequence availability will be applicable to limited numbers of species (unless a dramatic technological advance occurs). At the gross level, cytogenetic cross-species *in situ* hybridization (zoo-FISH) may be carried out. Single-chromosome DNA libraries can be constructed and used as fluorescent 'paints' to probe chromosomes of other species. A very convincing study of this sort has explored chromosome rearrangements in primates (Wienberg and Stanyon, 1995).

2.2 Processes of Chromosome Evolution

Chromosome evolution has an added dimension over DNA sequence evolution. There are constraints on gene order (Farr and Goodfellow, 1991). Some groups of genes are controlled together and need to be physically adjacent. However, the location of the majority of genes is unimportant and blocks of DNA can be shuffled with impunity both within and between chromosomes (Nadeau and Sankoff, 1997).

Telomeres are important structures that prevent loss of DNA from chromosome ends. When the ends of chromosomes fuse, remnants of telomeres are left, which explains why telomeric-like sequences can be found within chromosomes. Centromeres seem to be less important structures than telomeres, and their presence is variable.

Selective pressures on gene order are less than on gene sequence, so order may be a useful line of evidence for the elucidation of species divergence. Gene order in organelle genomes has indeed been used for phylogenetic inference (Sankoff *et al.*, 1990). To perform such inference it is necessary to have an understanding of the processes of chromosome evolution.

Visible changes in chromosomes arise frequently and are called chromosome aberrations. The changes can involve chromosome parts, whole chromosomes, or sets of chromosomes. Changes in chromosome structure are classified as:

- Deletions
- Duplications
- Inversions
- Translocations

Any aberration is a group of chromosome segments joined together according to these basic processes. The position of centromeres is subject to change. When aberrations become fixed in a population they lead to the evolution of a new stable karyotype. An evolutionary model can be constructed which describes the possible paths of transformation of one karyotype into another over time. The data are the observed chromosome segments which correspond in the different karyotypes. Segments may be characterized by their banding patterns, and parsimony has been used to choose a minimal set of steps to change one karyotype into another. Examples of applications are to the sibling species of *Drosophila* (Carson and Kaneshiro, 1976) and to primates (Yunis and Prakash, 1982).

Chromosome evolution at the cytological level has been reviewed by White (1973). Online Cytogenetics of Animals (OCOA) is compiled by F.W. Nicholas, S.C. Brown and P.R. Le Tissier. OCOA is a bibliographic resource which contains references to the literature on the cytogenetics of animals (not images of karyotypes, either normal or abnormal, at present). It is available at http://www.angis.su.oz.au/Databases/BIRX/ocoa/ocoa_form.html.

Maps of genes in vertebrates have been compiled by O'Brien (1987). Here we will describe the maps that are continually updated and available electronically.

3   VALUE OF COMPARATIVE MAPPING

Conservation of gene order in mammals is considerable. Establishing a gene order in one species makes it easier to establish the order in another if large blocks are conserved. The mouse is the major model organism for genetic purposes, being easily bred for experimentation including genetic manipulation. Such work is of enormous value to medical science and it would be hard to exploit the results of the human genome project without an equal effort on the mouse. The mouse is a suitable model for molecular studies. It is poorer for physiological studies and useless for surgical work or as a source of donor organs. The rat is favoured for physiological work and there are efforts to map the rat genome. For testing surgical procedures or as a source of donor organs, the pig is favoured because it is similar in size to humans.

Domestic livestock are being studied intensively in order to map traits of economic value. Cows, sheep and pigs are the major targets. Many economically important traits such as body-weight, amount of fat and susceptibility to disease are probably polygenic. They are called quantitative trait loci (QTLs) and, if a few loci dominate, the effect can be mapped genetically. Once linkage is established, the search for candidate genes can be made. QTLs are likely to be conserved across mammals and will add to our knowledge of abnormal human growth and development.

There are already examples where traits or disorders have been studied first in animals and then in humans. A well known example is the identification of a mutation in the ryanodine receptor gene (*RYR1*) as the cause of malignant hyperthermia which was described first in pigs (Fujii *et al.*, 1991) and second in humans (Gillard *et al.*, 1991). Another example is the Booroola fecundity gene in sheep, which may have a human homologue on chromosome 4q (Montgomery *et al.*, 1993).

4 SINGLE-SPECIES DATABASES

4.1 Primates

The main database for the order of genes in humans is the Genome Data Base compiled at Johns Hopkins University, Baltimore, Maryland (Fasman *et al.*, 1997). It is accessible at http://gdbwww.gdb.org/ as well as at a number of mirror sites around the world. Information about human phenotypes is contained in Online Mendelian Inheritance in Man (OMIM) at http://ww3.ncbi.nlm.nih.gov/Omim/searchomim.html.

GDB uses the term Genomic Segment to denote a region of the genome. Genomic Segments can range in size from points to regions as large as an entire chromosome. At present, GDB comprises descriptions of the following types of objects:

■   Genomic Segments, including genes, clones, amplimers (polymerase chain reaction (PCR) markers), breakpoints, cytogenetic markers, fragile sites, ESTs, syndromic regions, contigs and repeats.

■ Maps of the human genome, including cytogenetic maps, linkage maps, radiation hybrid maps, content contig maps and integrated maps. These maps can be displayed graphically via the Web.

■ Variations within the human genome including mutations and polymorphisms, plus allele frequency data.

Gene nomenclature is curated in GDB by the HUGO Nomenclature Committee, which assigns authoritative names. All objects in GDB can have any number of names or aliases, and may be retrieved using any one of them. Genes may have links to homologous gene entries in other species-specific databases. Currently these links are primarily to mouse genes in the Mouse Genome Database (MGD). Genes may also have links to protein sequence and structure database entries, to mutation information in locus-specific and genome-wide mutation databases, and to phenotype descriptions in OMIM and other phenotype databases. Any Genomic Segment may have links to entries in the sequence databases, and to any World Wide Web page or database containing additional information about the segment.

Maps that cover a region of interest can be retrieved and then displayed graphically, either singly or aligned to each other, using the Mapview program. GDB currently contains the following map types:

■ Cytogenetic maps
■ Linkage maps

These include whole genome maps from Genethon and CHLC, and other maps covering whole chromosomes or smaller regions.

■ Radiation hybrid maps

These include whole genome maps from the Stanford Human Genome Center and the Whitehead Institute.

■ Contig maps

These currently include the Whitehead Yeast Artificial Chromosome (YAC)–contig maps and a number of maps of specific regions.

■ Integrated maps

These include the RH Consortium Transcript maps, some integrated maps produced by chromosome committees, and some regional maps.

Outside GDB, other integrated human maps include CPROP and the Location Database (LDB). CPROP maps are produced by a program which uses reasoning with constraints. The maps produced are in postscript and are available from http://gdbdoc.gdb.org/letovsky/cprop/human/maps.html. LDB (Collins *et al.*, 1996) contains integrated maps made by an analytical method available from http://cedar.genetics.soton.ac.uk/public_html. Totally accurate and complete maps of the whole human genome will become available when the genome has been sequenced.

Work on gene mapping in non-human primates is not very extensive. The single species do not have their own databases. There is comparative information for 20 primates in the Mouse Genome Database (MGD) from the Jackson Laboratory at http://www.informatics.jax.org/homology.html.

4.2 Rodents

4.2.1 Mouse Genome Database (MGD)

MGD is a comprehensive resource for information on the biology and genetics of the laboratory mouse (Blake *et al.*, 1997). MGD contains the following kinds of information:

- References supporting all data in MGD.
- Gene, DNA marker, quantitative trait locus (QTL) and cytogenetic marker descriptions.
- Mouse genetic phenotypes, genetic interrelationships and polymorphic loci.
- Mammalian homology data.
- Genomic Segments (probes, clones, primers, ESTs, YACs, Massachusetts Institute of Technology (MIT) primers).
- Genetic and physical mapping data.
- Polymorphic loci related to specified strains.
- Information on inbred strains.

MGD provides links to relevant information in external databases wherever possible. Through homologies, MGD links to GDB, PigBase, SheepBase, BovMap and Ratmap. Through molecular probes and segments records, MGD links to the sequence databases: GenBank, GSDB, Entrez, DDBJ and dbEST. Through references, MGD links to Medline.

MGD contains information on mouse genes, DNA segments, cytogenetic markers and QTLs. Each record may include marker symbol, name, nomenclature history, alleles, aliases, chromosomal assignment, centimorgan location, cytogenetic band, EC number (for enzymes), phenotypic classifications, MGI accession numbers and supporting references.

The phenotype records include marker symbol, name, chromosomal assignment, phenotypic classifications, visible phenotype descriptions, information on mutations, gene interrelationships, information about related human diseases and a list of supporting references.

MGD contains homology information for mouse, human and over 60 other mammalian species. In addition to homology records, MGD provides homology information in a derivative of the Oxford Grid format. An Oxford Grid query form enables two species to be selected for comparison and the retrieval of a grid that provides an overall picture of homology between the selected species and links to homology records (Figure 5.2).

MGD contains information on probes, clones, primers, ESTs, YACs and sequence tagged sites (STSs). Information may include the origin of the probe/segment, sequence, vector, hybridizing loci, recombinant fragment length polymorphisms (RFLPs) and references.

In addition, information on genetic polymorphisms is extracted from probe/segment records in MGD. The Polymorphisms query form enables one to search directly for such information and display it.

MGD contains genetic mapping and linkage data, including haplotype data for linkage crosses, recombinant inbred (RI) strain distribution patterns, *in situ* hybridization data, deletion mapping information, translocations breakpoint mapping, somatic cell hybrids, concordance tables, congenic strains information and physical mapping information.

MIT is in the process of creating a physical map of the mouse by screening MIT primers against a large YAC library and building a contig map of the mouse genome. The goal is to include 10 000 markers (roughly 7000 MIT primers and 300 STSs).

Data from the major DNA mapping panels have been downloaded into MGD and are accessible through the DNA Mapping Panel Data Sets from MGD query form and from the Linkage Maps form. DNA Mapping Panel data may be displayed in tabular format, where each column represents a single offspring of the cross, and each row indicates for each locus which allele is present in each of the offspring. The order of rows is determined by linkage on the chromosome, and the locus nearest the centromere is at the top of the display. Centimorgan locations for loci in the cross are determined by the provider of the cross.

Marker records in MGD include available information on genetic mapping locations from several sources (MGD, Chromosome Committees, MIT). MGD provides mapping tools to generate graphical displays of genetic maps. Centimorgan locations are specific to the cross and may differ from estimates for the same loci found in MGD marker records.

MGD stores cytogenetic band information for many markers. The Cytogenetic Maps query form provides access to map images.

As of June 1997, MGD stores physical mapping data from MIT's mouse physical mapping project in which MIT primers and STSs are mapped against a mouse YAC library to build a contig map of the mouse. MGD provides query access to the MIT primers, STSs and YACs. In addition, MGD provides integrated genetic/physical displays in Web Map format. These map files are updated when the MIT mapping physical mapping dataset is updated.

MGD is easily accessed using query forms. Each query form is related to a particular kind of information in the database (references, marker information,

MGD I GXD I Encyclo I News I User Support I Docs I Mirrors
MGD Reports I Chr Comm I Nomen I Strains
Refs I Markers I Molecular I Homology I Mapping I GXD Index I Polymorphism I AccID

Each cell in the **Oxford Grid** represents a comparison of two chromosomes, one from each of the selected species. The number of homologies appears inside each colored cell and the color indicates a range in the number: Grey (1), Blue (2-10), Green (11-25), Orange (26-50), Yellow (50+). **Click on a colored cell to retrieve homology details.** A note about printing an Oxford grid...

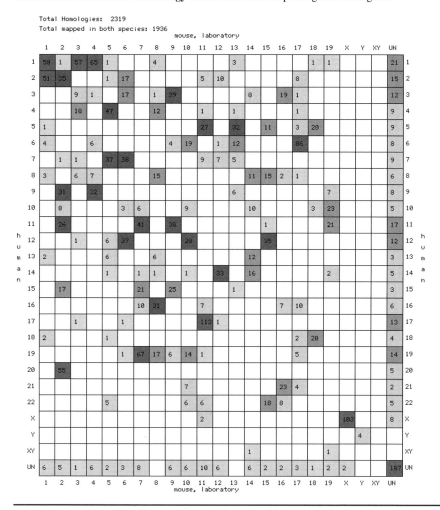

Figure 5.2   *A grid display from MGD showing the number of homologies recorded for each chromosome between human and mouse.*

homologies, molecular probes and segments, or mouse mapping data). The user selects a form, enters information in the form fields to build a query, and executes the query. If there are records matching the search criteria, a summary list of query results is displayed. Each item in the list contains some highlighted text denoting a hypertext link to detailed information, in the database. By selecting the link, the user can display the related information, which will also contain links to yet other information.

MGD is integrated with the Encyclopedia of the Mouse Genome, a genetic map viewing application. The browser may be configured to recognize the Encyclopedia as a 'helper' application. The Linkage Maps query form generates a linkage map in 'Encyclopedia' format. The Encyclopedia will automatically start up when the search is executed. The resulting map displays marker symbols arranged along the chromosome. Click on a marker and the browser will search MGD and retrieve the detail record for the selected marker for display in a browser window.

Another format for map display is the 'Web Map', a type of html file that provides a graphic display of cytogenetic and linkage maps on a WWW browser page. The appearance is similar to that of a 'PostScript' map. The marker symbols on the map are linked to marker records in MGD.

Phenotype records in MGD have their origin in the Mouse Locus Catalog (MLC), formerly an independent database. One can search for phenotype records using the Genes, Markers and Phenotypes query form. In addition, an integrated OMIM/MLC query form is available which allows one to enter one search string to execute a query in both MLC and OMIM.

MGD is closely linked with the GXD Index. The GXD Index is a collection of references to the scientific literature reporting data on endogenous gene expression during mouse embryonic development. MGD provides links to GXD Index records through genetic marker detail records. The GXD Index provides links to MGD references and genetic marker records.

4.2.2   European Collaborative Interspecific Backcross

The European Collaborative Interspecific Backcross (EUCIB) provides the resources for high-resolution genetic mapping of the mouse genome. This project has been being carried out at two centres: the Human Genome Mapping Project (HGMP) Resource Centre, UK, and the Pasteur Institute, France. A 1000-animal interspecific backcross between C57BL/6 and Mus spretus has been completed and DNAs prepared. Each backcross progeny mouse has been scored for three or four markers per chromosome, completing an anchor map of 70 loci across the mouse genome. A 1000-animal cross provides a genetic resolution of 0.3 cM with 95% confidence. Completion of the anchor map allows the identification of pools of animals recombinant in individual chromosome regions and allows a rapid two-stage hierarchical mapping of new loci. New markers are first analysed through a panel of 40 to 50 mice in order to identify linkage to a chromosome region. Subsequently, the new marker is analysed through a panel of mice identified as carrying recombinants within that chromosome region.

The backcross is supported by a database called MBx which stores mouse, locus and probe data. It stores all allele data at each chromosome locus for each of the 1000 backcross progeny. Allele data are presented as a scrollable matrix on screen. When a new marker is analysed through the backcross, MBx provides lod score information to indicate linkage to a chromosome region. Using the computer screen, recombinant mice in this chromosome region can be selected for the second stage of hierarchical screening. In addition, at each stage, MBx will not only calculate the available lod scores for closely linked markers but will also determine genetic order with respect to closely linked markers by minimizing the number of recombinants.

MBx has been extended to hold physical mapping data. For each physical marker, details of the marker itself, primers used to screen YAC libraries and positive YAC clones are stored. Physical data can be exported to an external contig building program called SAM (system for assembling markers) which produces partially ordered physical maps with alignment of YAC clones. These maps can be further manipulated via a graphical interface and the resulting solution saved to the database.

EUCIB data including physical and genetic maps are available via the World Wide Web at http://www.hgmp.mrc.ac.uk/MBx/.

4.2.3   RATMAP

RATMAP (http://ratmap.gen.gu.se/) is a database covering genes physically mapped to chromosomes in the laboratory rat. It is maintained by the Department of Genetics in the University of Goteborg. In addition to chromosomal localization of genes and anonymous DNA sequences, RATMAP gives information about methods and accuracy of localization and a short description of the genes. Appropriate literature references to each mapped rat gene are also given. Furthermore, RATMAP supports comparative mapping to several species: bovine, human and mouse.

4.3   Artiodactyls

4.3.1   The ARK System

The ARK family of animal genome mapping databases is being developed at the Roslin Institute, Edinburgh (http://www.ri.bbsrc.ac.uk). It is currently being used for cattle, pig, sheep and chicken. The interface is divided into sections depending on what kind of objects being queried (loci/markers, publication references, clones and libraries). Each section has a form allowing queries to that part of the database. The various sections of the database are cross-linked, allowing related references to be followed. Central to the ARK schema is the abstracted concept of an experiment, which ties loci, references, experimental details and results together. This allows for a much richer and more detailed storage and representation of data. New data currently in the database include experimental method details, probes, clones and libraries. This front-end is coupled with the Anubis genome viewer for graphical map representations.

4.3.2   Cattle Genome Database

The Cattle Genome Database (CGD) is intended as a fast access database for information on the genome of cattle (http://spinal.tap.csiro.au/cgd.html, soon to be replaced by http://spinal.tag.csiro.au/cgd.html). It grew out of the Cattle Genotypic Database which was designed to construct a genetic linkage map of the bovine genome (Barendse *et al.*, 1994, 1997). In addition to the genetic maps of chromosomes, information on loci, contact addresses, citations and comparative information will also be contained in this database.

There are currently more than 36 laboratories that have contributed to the CGD. Contributions to the Database are made through the genotyping of the International Bovine Reference Panel, which consists of full-sib cattle families from a wide range of sources. Genotypes are collected and linkage analyses processed at the Molecular Animal Genetics Centre. Regular updates of this map are circulated to contributors. With the publication of the latest map (Barendse *et al.*, 1997), it was decided that this resource should be made widely available to the gene mapping community. The public map on the web site is updated regularly. This information is edited by Dr W. Barendse at the CSIRO Molecular Animal Genetics Centre, Brisbane, Australia.

4.3.3   Further Resources for Cattle

The Animal Genome Database in Japan has cytogenetic maps for cattle (as well as for pig, mouse, human, rat and sheep) (http://ws4.niai.affrc.go.jp/). BOVMAP is compiled at the INRA Biotechnology Laboratories, Jouy-en-Josas, France (http://locus.jouy.inra.fr/). Cattle maps are held by the Meat Animal Research Center, Clay Center, Nebraska (http://sol.marc.usda.gov/genome/cattle/cattle.html). BOVGBASE (using an earlier version of the ARK software) is referenced from Texas A and M University (http://bos.cvm.tamu.edu/bovgbase.html).

4.3.4   PiGBASE and Related Resources

PiGBASE is implemented in the ARK system at the Roslin Institute (Figure 5.3). The PiGMaP project involved 17 European laboratories and placed 245

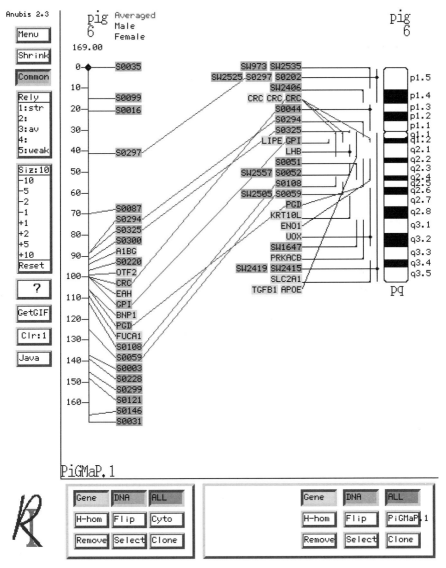

Figure 5.3 *Comparison of the linkage and cytogenetic maps of chromosome 6 from PiGBASE. (Reproduced with permission from Alan Archibald.)*

polymorphic genetic markers on the pig genome to give a resolution of at least 20 cM over most regions.

Informative reference populations have been established using a Chinese Meishan and European Large White cross and diverse crosses between Wild Boar and various European commercial breeds. The F2 individuals are highly informative. In addition, 130 loci have been mapped cytogenetically by *in situ* hybridization. PiGBASE contains the information resulting from the PiGMaP project and is being extended to include other types of data (e.g. radiation hybrid mapping).

The US Pig Gene Mapping website is at Iowa State University (http://www.public.iastate.edu/~pigmap/pigmap.html). Pig maps are also held by the Meat Animal Research Center, Clay Center, Nebraska (http://sol.marc.usda.gov/genome/swine/swine.html).

4.3.5   SheepBase

The purpose of SheepBase is to provide an up-to-date compilation of published data from sheep genome mapping projects. Information is presented using the WWW interface, where both physical and linkage maps of the sheep genome are available, together with information on individual loci and associated references.

SheepBase data are maintained by an international editorial committee and held at AgResearch, Invermay Agricultural Centre, Dunedin, New Zealand (http://dirk.invermay.cri.nz/).

Information has been entered into the database to represent best the published data available. In so doing, not all information can be represented completely. The cytogenetic map contains some loci that have been mapped by linkage to a chromosome, but for which no map position is available. Loci mapped first by somatic cell hybrids, and later included on the linkage map, appear on the linkage map only. The cytogenetic anchoring of the linkage groups to particular chromosomes is based on one or more markers in the linkage group being assigned to the chromosome by one of three methods:

- Direct *in situ* hybridization.
- Somatic cell hybrid analysis.
- Assignment to a cytogenetically equivalent bovine chromosome.

The genetic distance between loci is given in centimorgans, and its calculation is based on the Kosambi mapping function (if not otherwise indicated). The orientation of the linkage maps is not specified.

Further sheep mapping information is available from the Centre for Animal Biotechnology, Melbourne, Australia (http://rubens.its.unimelb.edu.au/~jillm/jill.html).

4.4   Carnivores

4.4.1   DogMap

DogMap (http://ubeclu.unibe.ch/itz/dogmap.html) is an international collaboration between 42 laboratories from 20 different countries towards a low-resolution canine marker map under the auspices of the International Society for Animal Genetics (ISAG).

The map under development should achieve a resolution of about 20 cM and some of the markers should be mapped physically. The participants have agreed to use microsatellites as markers on a common panel of reference families which will provide the backbone of the marker map. It is foreseen also to include type I markers in the mapping effort and to produce cosmid-derived microsatellites for physical mapping. For this purpose part of the effort focuses on the standardization of the canine karyotype. Special attention is given to hereditary diseases, and efforts are under way to establish resource families either by collecting families or by specific breeding.

A point of emphasis of the DogMap project is the setting up of an internationally accessible database for handling the mapping data based on the BovMap software. This is maintained by Etienne Baumle and Gaudenz Dolf of the Institute of Animal Breeding, University of Berne, Switzerland.

4.4.2   Dog Genome Project

The Dog Genome Project is a collaborative study involving scientists at the University of California, the University of Oregon and the Fred Hutchinson Cancer Research Center, aimed at producing a map of all dog chromosome. Dogs vary greatly in size, shape and behaviour, and interesting insights are expected to be forthcoming. The information is at http://mendel.berkeley.edu/dog.html.

A major goal of the Dog Genome Project is to develop a map that will be useful to the entire scientific community for the purpose of mapping genes causing inherited diseases in dogs. These diseases include cancer, epilepsy, retinal degeneration, bleeding disorders and skeletal malformations. The map being produced by the Dog Genome Project will be of great assistance to veterinary medicine. It will also allow more effective breeding practices to eliminate many genetic diseases from breeds currently afflicted.

4.5   Marsupials and Monotremes

The mapping of human X genes in marsupials and monotremes has proved informative about evolutionary history. Markers on human Xq and the pericentric region are located on X in marsupials and monotremes. Markers distal to Xp11.3 are autosomal in both marsupials and monotremes. Human Xq is therefore the ancestral X and much of Xp must have been added from the autosomes in early eutherians. A table of these and other mapping results is available at http://www.latrobe.edu.au/genetics/roobase.html.

4.6   Chicken

Chicken genome mapping is being carried out as an international collaboration using two experimental crosses, East Lansing from the USA and Compton from the UK, as reference populations. The main web sites are at Roslin (http://www.ri.bbsrc.ac.uk/chickmap/) and East Lansing (http://poultry.mph.msu.edu/). The data are contained in ChickBase which is based on the ARK software. The genetic map contains over 1100 loci and is nearing completion. Many genes have been mapped and it is estimated that there are 170 conserved segments and 150 rearrangements between humans and chicken.

4.7   Teleost fish

4.7.1   *Tilapia Genome Project*

A first-generation genetic map for the fish tilapia is being constructed at the University of Hawaii (http://tilapia.unh.edu/WWWPages/TGP.html). This is a group of fish important to aquaculture around the world. Microsatellite markers consisting of a variable number of dinucleotide repeats have been isolated from an enriched genomic DNA library. These markers will be useful for characterizing the genomic composition and level of inbreeding of commercial tilapia strains, many of which have been derived by hybridization of several related species of *Oreochromis*. The genetic map will facilitate the improvement of strains with respect to traits of commercial importance, such as growth rate and flesh quality, through marker-assisted selection.

4.7.2   *Zebrafish Genome Project*

The zebrafish linkage map contains 652 markers genotyped in one or both of two haploid mapping panels (Johnson *et al.*, 1996). Information on zebrafish is available at http://zfish.uoregon.edu/(The Fish Net).

4.7.3   *Fugu Genome Scanning*

Fugu is of interest because it has a genome one-eighth the size of that of the human with a conserved exon/intron structure (Elgar *et al.* 1993). It is therefore a useful model to work out human gene structure with a reduced sequencing effort (compared with, for example, the mouse). The Fugu Genome Project does not involve the construction of a conventional map but is sample sequencing from 1000 cosmid clones. This will provide information about local conservation of synteny between fugu and human, as well as discovering several thousand fugu genes. Information about this project is available at http://fugu.hgmp.mrc.ac.uk/.

4.8   Online Mendelian Inheritance in Animals

Online Mendelian Inheritance in Animals (OMIA; (http://www. angis.su.oz.au/Databases/BIRX/omia) is compiled by F.W. Nicholas, S. C. Brown and P. R. Le Tissier. OMIA is a database of the genes and phenes that have been documented in a wide range of animal species other than those for which databases already exist (human, rat and mouse). A phene is a word or words that identify a familial trait. For single-locus traits, the words correspond to one of the phenotypes that arise from segregation at that locus. OMIA also includes multifactorial traits and disorders. OMIA is modelled on, and is complementary to, McKusick's Online Mendelian Inheritance in Man (OMIM).

OMIA has been under construction since 1980. It contains references to publications on any trait or disorder for which familial inheritance has been claimed, except for a gap for the years 1982–1988, which is gradually being filled. It does not contain sequence data. Each reference has been indexed to one or more phenes and/or genes, and to species. At present, an entry consists solely of references arranged chronologically for each phene in each species.

5   PROTEIN FAMILIES

5.1   Homologous Vertebrate Genes (HOVERGEN)

In Section 1.3 the importance of complete knowledge of the members of a gene family in an individual in understanding between species relationships of genes was explained. HOVERGEN attempts to compile such information for vertebrates (Duret *et al.*, 1994) and allows one easily to select sets of homologous genes among a given set of vertebrate species. HOVERGEN gives an overview of what is known about a particular gene family and contains the nuclear and mitochondrial vertebrate sequences subset of GenBank. Coding sequences have been classified into gene families. Protein multiple alignments and phylogenetic trees have been calculated for each family.

A graphical interface has been developed to visualize and edit trees. Genes are displayed in colour, according to their taxonomy. Users have directly access to attached information or multiple alignments by clicking on genes. This graphical tool gives thus a rapid and simple access to all data necessary to interpret homology relationships between genes: phylogenetic trees of gene families, taxonomy, GenBank information and protein multiple alignments.

5.2   Other Databases

GDB is experimenting with ways of presenting gene family information in the Gene Family Database (http://gdbdoc.gdb.org/~avoltz/home.html). The gene families presently listed are PAX, cytochrome P450, insulin and insulin-like growth factors, amyloid precursor protein and DNA repair proteins.

The Homeobox Page is maintained by Thomas R. Burglin at the University of Basel (http://copan.bioz.unibas.ch/homeo.html).

5.3   Mendel

Mendel is a database of plant genes but the methodology used in constructing it may be of interest to workers on vertebrates (Beckstrom-Sternberg *et al.*, 1997). The purpose of this database is to provide a common system of nomenclature for substantially similar genes across the plant kingdom. A description of the various fields in Mendel follows:

- GeneFamily – mnemonic designation for plant-wide gene family.
- ISPMB_Number – ID number for plant-wide gene family.
- EC_Number – Enzyme Commission number.
- GPN – numeric designation for plant-wide gene family.
- GeneProduct – product of the gene.
- GeneProductAbbrev – product of the gene, abbreviated.
- WorkingGroup – group responsible for designations within this category of genes and organizer of the group.
- MendelNo – ID number for individual gene.
- Species – genus and species = OS field.
- MemberNo – sequential number for membership in multigene family within single species.
- GeneSynonym – other symbols that have been employed for this specific gene.
- DNAseqAC – three fields follow this tag: 1. accession number (AC) in GenBank, EMBL databases, etc.; 2. clone type; 3. Synonym.
- EPD – accession number in Eukaryotic Promoter Database.

■ Genome – type of genome in which gene occurs: this field occurs only when the genome is not either plastid, or mito (mitochondrial). Nuclear is the unstated default.

■ Expression_conditions – conditions of expression of the gene.

6   COMPARATIVE MAPPING DATABASES

6.1   Human/Mouse Homologies

As described in Section 4.2.1, the Mouse Genome Database contains a considerable amount of comparative information. There are also two other lists of human–mouse homologies. One is at the MRC Mammalian Genetics Unit (http://www.mgu.har.mrc.ac.uk/mgu/homolog.html) and this list is published annually in Mouse Genome. The other is at the NCBI (http://www3.ncbi.nlm.nih.gov/Homology/) and was published as DeBry and Seldin (1996).

6.2   Development of a Vertebrate Comparative Mapping Database

Numerous phenotypes have been mapped in vertebrates that can contribute to comparative mapping between similar species. However, to be successful across the entire range of vertebrates, the requirements are:

■ Accurate single species gene orders as described in Section 4.
■ An accurate compilation of protein families as described in Section 5.

The single-species databases can provide the mapping information by links from the comparative database. The heavy work is the estimation of the protein homologies. For n objects there are $n(n - 1)/2$ pair-wise comparisons that can be made. The protein family database has to be built up carefully and kept under constant review. The comparative maps can be built on the fly to reflect the latest protein homology and gene order information. A possible implementation is The Comparative Animal Genome Database (TCAGDB), being constructed at Roslin.

ACKNOWLEDGEMENTS

A large amount of useful information has been obtained and summarized from the various web sites listed in this chapter. The author is grateful to all those who have made this information publicly available in an attractive and accessible manner. The originators of this information are listed on the respective web sites.

REFERENCES

Aparicio S, Hawker K, Cottage A, Mikawa Y, Zuo L, Venkatesh B, Chen E, Krumlauff R, and Brenner S (1997) Organisation of the *Fugu rubripes Hox* clusters: evidence for continuing evolution of vertebrate *Hox* complexes. *Nature Genet.* **16**, 79–83.

Barendse W, Armitage SM, Kossarek LM, Shalom A, Kirkpatrick BW, Ryan AM, Clayton D, Li L, Neibergs HL, Zhang N, Grosse WM, Weiss J, Creighton P, McCarthy F, Ron M, Teale AJ, Fries R, McGraw RA, Moore SS, Georges M, Soller M, Womack JE and Hetzel DJS (1994) A genetic linkage map of the bovine genome. *Nature Genet.* **6**, 227–235.

Barendse W, Vaiman D, Kemp SJ, Sugimoto Y, Armitage S, Williams JL, Sun HS, Eggen A, Agaba M, Aleyasin SA, Band M, Bishop MD, Buitkamp J, Byrne K, Collins F, Cooper L, Coppettiers W, Denys B, Drinkwater RD, Easterday K, Elduque C, Ennis S, Erhardt G, Ferretti L, Flavin N, Gao Q, Georges M, Gurung R, Harlizius B, Hawkins G, Hetzel DJS, Hirano T, Hulme D, Joergensen C, Kessler M, Kirkpatrick BW, Konfortov B, Kostia S, Kuhn C, Lenstra J, Leveziel H, Lewin HA, Leyhe B, Li L, Martin Buriel I, McGraw RA, Miller JR, Moody DE, Moore SS, Nakane S, Nijman I, Olsaker I, Pomp D, Rando A, Ron M, Shalom A, Soller M, Teale AJ, Thieven I, Urquhart B, Vage D-I, Van de Weghe A, Varvio S, Velmala R, Vilkki J, Weikard R, Woodside C, Womack JE, Zanotti M and Zaragoza P (1997) A medium density genetic linkage map of the bovine genome. *Mammal. Genome* **8**, 21–28.

Bassett DE, Boguski MS, Spencer F, Reeves R, Goebl M and Hieter P (1995) Comparative genomics, genome cross-referencing and XREFdb. *Trends Genet.* **11**, 372–373.

Beckstrom-Sternberg SM, Reardon E and Price C (1997) *The Mendel Database.* http://probe.nalusda.gov:8300/cgi-bin/browse/mendel.

Blake JA, Richardson JE, Davisson MT, Eppig JT, Battles J, Begley D, Blackburn R, Bradt D, Colby G, Corbani L, Davis G, Doolittle D, Drake T, Frazer K, Gilbert J, Grant P, Lennon-Pierce M, Maltais L, May M, McIntire M, Merriam J, Ormsby J, Palazola R, Ringwals M, Rockwood S, Sharpe S, Shaw D, Reed D, Stanley-Walls M and Taylor L (1997) The Mouse Genome Database (MGD) a comprehensive public resource of genetic phenotypic and genomic data. *Nucleic Acids Res.* **25**, 85–91.

Carson HL and Kaneshiro KY (1976) *Drosophila* of Hawaii: systematics and ecological genetics. *Ann. Rev. Ecol. Systematics* **7**, 311–345.

Collins A, Frezal J, Teague J and Morton N (1996) A metric map of humans: 23 500 loci in 850 bands. *Proc. Natl. Acad. Sci. U.S.A.* **93**, 14 771–14 775.

Comparative Genome Organization (1996) First International Workshop *Mammal. Genome* **7**, 717–734.

DeBry RW and Seldin MF (1996) Human/mouse homology relationships. *Genomics* **33**, 337–351.

Duret L, Mouchiroud D and Gouy M (1994) HOVERGEN a database of homologous vertebrate genes. *Nucleic Acids Res.* **22**, 2360–2365.

Elgar G, Brenner S, Sandford R, Macrae A, Venkatesh B and Aparicio S (1993) Characterization of the pufferfish (Fugu) genome as a compact model vertebrate genome. *Nature* **366**, 265–268.

Farr CJ and Goodfellow PN (1991) Hidden messages in genetic maps. *Science* **258**, 49.

Fasman KH, Letovsky SI, Li P, Cottingham RW and Kingsbury DT (1997) The GDB™ Human Genome Database Anno 1997. *Nucleic Acids Res.* **25**, 72–80.

Fujii J, Otsu K, Zorzato F, de Leon S, Khanna VK, Weiler JE, O'Brien PJ and MacLennan DH (1991) Identification of a mutation in the porcine ryanodine receptor associated with malignant hyperthermia. *Science* **253**, 448–451.

Gillard EF, Otsu K, Fujii J, Khanna VK, de Leon S, Derdemezi J, Brit BA, Duff CL, Worton RG and MacLennan DH (1991) A substitution of cysteine for arginine 614 in the ryanodine receptor is potentially causative of human malignant hyperthermia. *Genomics* **11**, 751–755.

Green P, Lipman DJ, Hillier L, States D, Waterston R and Claverie J-M (1993) Ancient conserved regions in new gene sequences and the protein databases. *Science* **259**, 1711–1716.

Johnson SL, Gates MA, Johnson M, Talbot WS, Horne S, Baik K, Rude S, Wong JR and Postlethwait JH (1996) Centromere-linkage analysis and the consolidation of the zebrafish genetic map. *Genetics* **142**, 1277–1288.

Lundin LG (1993) Evolution of the vertebrate genome as reflected in paralogous chromosome regions in man and the house mouse. *Genomics* **16**, 1–19.

Miyamoto MM (1996) A congruence study for molecular and morphological data for eutherian mammals. *Mol. Phylogenet. Evol.* **6**, 373–390.

Montgomery GW, Crawford AM, Penty JM, Dodds KG, Ede AJ, Henry HM, Pierson CA, Lord EA, Galloway SM, Schmack AE, Sise J-A, Swarbridk PA, Hanrahan V, Buchanan FC and Hill DF (1993) The ovine Booroola fecundity gene (*FecB*) is linked to markers from a region of human chromosome 4q. *Nature Genet.* **4**, 410–414.

Nadeau JH and Sankoff D (1997) Landmarks in the Rosetta Stone of mammalian comparative maps. Nature Genet. **15**, 6–7.

O'Brien SJ (ed.) (1987) *Genetic Maps*. Cold Spring Harbor Laboratory New York.

Sankoff D, Cedergren R and Abel Y (1990) Genomic divergence through gene rearrangements. *Methods Enzymol.* **183**, 428–438.

Smith MW and Doolittle RF (1992) A comparison of the evolutionary rates of the two major kinds of superoxide dismutase. *J. Mol. Evol.* **34** 175–184.

White MJD (1973) *Animal Cytology and Evolution*. Cambridge University Press Cambridge.

Wienberg J and Stanyon R (1995) Chromosome painting in mammals as an approach to comparative genomics. *Curr. Opin. Genet. Dev.* **5**, 724–733.

Yunis JJ and Prakash O (1982) The origin of man: a chromosomal pictorial legacy. *Science* **215**, 1525–1529.

6 Radiation Hybrid Mapping

Linda McCarthy[1] and Carol Soderlund[2]

[1] Department of Genetics, University of Cambridge, Downing Site, Tennis Court Road, Cambridge CB2 3EH and [2] The Sanger Centre, Wellcome Trust Genome Campus, Hinxton, Cambridge CB10 1SA, UK

1 INTRODUCTION

Radiation hybrid (RH) mapping is the most efficient method for generating long-range genomic maps using both polymorphic and non-polymorphic markers. There are a number of advantages to radiation hybrid mapping. Non-polymorphic markers are informative, making them particularly useful for anchoring clone contigs which frequently do not contain polymorphic markers. RHs offer high map resolution for relatively small numbers of hybrids. A high-resolution (100 kb) whole genome map can be generated using fewer than 100 hybrids. Apart from the efficiency of RH mapping as a stand-alone mapping tool, the additional strength of this technique lies in allowing the efficient integration of physical and genetic maps, by facilitating the resolution of cosegregating genetic markers lying in recombination cool-spots on the meiotic map, and anchoring Yeast Artificial Chromosome (YAC) contigs to the genome map.

Radiation hybrids are generated by irradiating the donor cell line with a lethal dose of X-rays to fragment the chromosomes. These cells are fused with the recipient cell line, using either Sendai virus or polyethylene glycol (PEG). DNA fragments from the donor cells are rescued by the recipient cells, both by integration of donor DNA into recipient chromosomes and by maintenance of recombinant donor chromosomes independently of the recipient chromosomes. The recipient cell line will contain a selectable marker; the most frequently used are thymidine kinase deficiency (TK⁻) or hypoxanthine phosphoribosyltransferase deficiency (HPRT⁻). Cells containing either of these markers will not grow in media containing HAT (hypoxanthine, aminopterin, thymidine). The only viable post-fusion cells are hybrids containing the donor wild-type TK or HPRT genes.

Guide to Human Genome Computing, 2nd edition   Copyright © 1998 Academic Press Limited
ISBN 0-12-102051-7

Non-selected chromosome fragments are also maintained in these hybrids. A RH panel incorporating at least 20% of the donor genome per hybrid is necessary for efficient map generation.

Most of the first radiation maps were constructed using hybrid panels carrying a single copy of a unique human chromosome. A disadvantage of this technique is that a different panel must be created for each chromosome. The original protocol of Goss and Harris (1975) used a diploid human fibroblast cell line as the DNA donor, instead of a somatic cell hybrid containing a single human chromosome on a rodent background. By using a whole genome radiation hybrid (WG-RH) panel, anonymous markers can be both assigned to a chromosome and localized on that chromosome.

The basic principle of RH mapping is that the closer two loci are on a chromosome, the less likely they are to be separated by a radiation-induced break. Therefore, markers that are closely linked show correlated coretention patterns across the hybrid panel, where markers located a large distance from one another are retained almost independently. WG-RH mapping has been reviewed extensively elsewhere (Walter and Goodfellow, 1993; Leach and O'Connell, 1995; McCarthy, 1996). The two most extensively used RH mapping programs are RHMAP (Boehnke *et al.*, 1991) and RHMAPPER (Slonim *et al.*, 1996). In this chapter we will demonstrate the analysis of two WG-RH datasets using these programs, discussing the merits and limitations of both packages. We will also briefly describe the RADMAP analysis package (Matise *et al.*, 1994).

2 BACKGROUND

The technology for physical map generation using irradiation and fusion gene transfer (IFGT) was first developed by Goss and Harris (1975). This technique was generally underexploited until advances in molecular genetics allowed efficient PCR screening of the RH panels. It was not until 15 years later, when Cox *et al.* (1990) used RHs for the construction of a map of a region of chromosome 21, that the method became widely used for mapping individual mammalian chromosomes (Burmeister *et al.*, 1991; Richard *et al.*, 1991, 1993; Frazer *et al.*, 1992; Winokur *et al.*, 1993; Kumlien *et al.*, 1994; James *et al.*, 1994; Raeymaekers *et al.*, 1995; Shaw *et al.*, 1995; Bouzyk *et al.*, 1996). These maps were constructed using hybrids generated by IFGT between a donor somatic cell hybrid containing a single human chromosome, and the recipient rodent cell line. Mapping the entire human genome using this approach is impractical, because a panel of 100–200 hybrids per chromosome would require screening 2000–4000 hybrids to generate a genomic map. Whole genome RH mapping was revived by Walter *et al.* (1994). Using a panel of 44 WG-RHs, an ordered map of the long arm of chromosome 14 was generated containing 40 markers, with five gaps in the map. These data suggested that a single panel of 100 hybrids could be used to generate a WG-RH map.

The technique has recently come of age with the publication of several RH maps of the human genome. To construct an RH map of the human genome, Gyapay *et al.* (1996) generated a panel of 168 RHs, which they used to produce a framework map of 404 microsatellite markers on which 374 expressed sequence tags (ESTs) were mapped. A subset of this panel containing 91 hybrids was used by Hudson *et al.* (1995) to map 6193 markers, which were a combination of both genetically mapped and unmapped sequence tagged sites (STSs), including markers that were not sufficiently polymorphic to be typed on the Genethon human genetic map. The latest data release from Hudson *et al.* reports 12 508 markers mapped on the human panel, as well as integrations of the genetic, RH and YAC contig maps (Whitehead Institute/MIT Centre for Genome Research[1]). Although both maps were generated using the same RH panel, the panel was grown up separately for each project. As there would be up to 25% difference in the donor DNA fragment content of each hybrid between individual 'grow-ups' of the panel (K. Schmitt, unpublished data), in practical terms the maps may as well have been generated using different panels. The panel used by Hudson *et al.* was grown commercially (Research Genetics[2]) on a large scale and marketed as Genebridge4 (GB4) for distribution to the scientific community. Stewart *et al.*, (1997) generated a high-resolution genome map using the Stanford G3 panel of 83 radiation hybrids, which is also commercially available (Research Genetics). This map consists of 10 478 STSs, of which 5049 are framework markers.

Through an international consortium, over 16 000 human genes have been mapped relative to a framework map comprising about 1000 polymorphic markers (Schuler *et al.*, 1996). The gene-based STSs were developed and mapped to the GB4 RH panel, the G3 RH panel and the CEPH YAC panel (Dausset *et al.*, 1992). Some 70% of the genetic markers were common to all three maps, and these were used to assemble the integrated gene map. The gene map is available on http://www.ncbi.nim.nih.gov/SCIENCE96/.

The program RHMAP was used by Gyapay *et al.* (1996) and RHMAPPER by Hudson *et al.* (1995). RADMAP was used to build the framework for the Schuler *et al.* (1996) gene map, with various laboratories in the consortium using RHMAP and RHMAPPER to place markers on the map.

3 RH MAPPING CONCEPTS

Definitions

Theta (θ) is the breakage frequency between two markers.

---

[1] Whitehead Institute/MIT Center for Genome Research, Data Release 11.
   http://www.genome.wi.mit.edu/ftp/distribution/human_STS_releases/oct96.
[2] Research Genetics, 2130 Memorial Parkway SW, Huntsville, AL 35801, USA.

D is the distance between two markers calculated by $-\ln(1 - \theta)$, used instead of θ because it is additive. D is expressed in Rays.

1 Ray = 100 cRays. A distance of 1 $cR_{(N)}$ between two markers corresponds to a 1% breakage frequency. These units are qualified by the X-irradiation dose (N) used in panel generation. Higher radiation doses generate shorter donor DNA fragments, leading to a higher map resolution.

Likelihood is the probability that the observed data were produced from a given map, where the map is an ordered set of markers with distances between them.

lod (likelihood of odds) is the difference of the $\log_{10}$ likelihoods of two orders; for example, if a lod score is 3.0, then the ratio of their likelihoods is 1000 : 1.

Pairwise lod score is the ratio of the likelihood of the two markers being linked, with the likelihood of them being totally unlinked (i.e. $\theta = 1$).

Retention frequency (RF) for a marker is the ratio of its positives to the total number of hybrids. The average RF is the retention for the panel.

Obligate breaks are the number of breaks that are obligatory in any given marker order. In other words, the number of times that a 0 is followed by a 1, or a 1 is followed by a 0. For example, the hybrid x = (00111000111) has three obligate breaks.

3.1 Framework and Comprehensive Maps

The techniques for building RH maps are similar to those for building genetic maps; in fact, some of the software programs for RH mapping are modifications of genetic mapping programs. There are two recognized classes of genetic maps: framework and comprehensive (Keats *et al.*, 1991; Matise *et al.*, 1994). Framework maps have low resolution and high locus support consisting of loci mapped at odds of 1000 : 1 or greater. Comprehensive maps have higher resolution and low local support as they consist of all syntenic markers placed uniquely in their most likely position. Framework and comprehensive maps, or a combination of the two, are also used for RH mapping. A placement map is a framework with additional markers binned; the binned markers between two framework markers are not ordered (or ordered only roughly). An example of a typical description of an RH map is 'The most likely radiation hybrid map localised the 95 STSs into 54 unique map positions, 34 with odds of 1000 : 1 or greater; the comprehensive map localises all but 17 STSs with odds exceeding 10 : 1' (Shaw *et al.*, 1995). Note that odds of 10 : 1 do not imply a poor placement for closely localized markers. For markers separated by few obligate breakpoints, a number of different orders are frequently generated with similar likelihoods. The two highest scoring orders will therefore have a small lod score.

3.2 Retention

A RH panel incorporating between 20% and 80% of the donor genome per hybrid is necessary for efficient map generation. Where the retention rate is less than 20%, the number of hybrids screened needs to be increased to compensate for the reduced genome coverage. Retention rates above 80% make mapping difficult, as most markers will be positive in these regions resulting in a loss of mapping power. Significant differences in retention of donor DNA fragments across the genome distort map distances in these regions. Donor fragments containing the selectable marker are always retained in the hybrids, resulting in an above average retention for this chromosome, increasing to 100% retention at the selectable marker. Sometimes centromeric, and to a lesser extent telomeric, regions are also retained at an above average rate. This appears to be an organism-specific effect.

A significant number of chromosomes in the human Genebridge4 panel display higher retention rates in pericentromeric and telomeric regions. These effects are in less evidence in the T31 mouse WG-RH panel (unpublished data). This effect is due to some donor DNA fragments not being incorporated into the recipient chromosomes, but maintained as recombinant donor chromosomes containing donor centromeric and telomeric regions. Highly retained regions can contain fewer obligate breakpoints, contracting the map distances in these regions. Flanking the selective marker, for example, the retention rate drops rapidly to the average for the panel; consequently there are a higher than average number of obligate breakpoints covering this physical distance. Map distance distortion is a reality in regions spanning spikes in retention rate, leading to contraction of map distances at plateaus of high retention, flanked by map expansion in regions of rapidly increasing or decreasing retention rate. This should be considered carefully when analysing datasets. Retention models are available in some software packages to accommodate this effect, two of which are the centromeric retention model and the general retention model. Retention models will be described more fully in Section 4.1.3 of this chapter.

3.3 Radiation Dosage

The eventual resolution of the panel depends on the initial radiation dose used to fragment the donor chromosomes. A higher dose generates shorter DNA fragments leading to a potentially higher resolution RH panel. The actual average fragment length of a panel can be truly estimated only when a panel is saturated with markers such that all fragment ends are revealed, and can then be compared with a high-resolution physical map. Clearly we are not yet in a position to make these calculations unambiguously for WG-RH panels. Nevertheless, estimates of average fragment size have been made for some RH panels. The GB4 panel was generated using 3000 rads of radiation from an X-ray tube source, and it is estimated that the average fragment size is approximately 10 Mb. The G3 panel was generated using 10 000 rads of radiation from a cobalt source, and its average fragment size has been estimated as 4 Mb. (The estimates of fragment lengths are taken from Schuler *et al.* (1996).) Experience shows that radiation doses in rads are not equivalent between laboratories. A third commercially available panel (Research Genetics), the TNG human WG-RH panel, was generated at the Stanford Human Genome Centre with 50 000 rads of radiation. Stewart *et al.* (1997) estimate that the three panels, GB4, G3 and TNG, can be used to order markers to 1 Mb, 240 kb and 50 kb resolution respectively.

3.4 Radiation Hybrid Ploidy

Most RH mapping software takes into account the ploidy of the hybrids. In diploid panels, a proportion of the donor DNA fragments will contain genomic segments in common. The DNA fragment 'overlaps' will disguise the internal fragment ends resulting in two obligate breakpoints instead of four, decreasing the resolution for this portion of the map (Figure 6.1). This effect is a function of the square of the retention frequency of a marker and will have a small distortion effect on the map distances, varying as the retention frequency varies (Walter *et al.*, 1994).

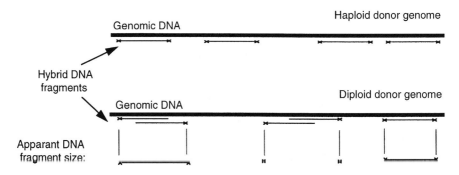

Figure 6.1 *Obligate breakpoints are represented by an x. Overlapping DNA fragments in the diploid hybrid disguise fragment ends. This results in fewer obligate breakpoints than would be contained in a haploid hybrid with similar retention rate and fragment size. The overall effect is a greater variation in fragment sizes in diploid hybrids, with some loss of mapping power. For this reason, it is important to specify the ploidy of a hybrid panel in any analysis program.*

4 RADIATION HYBRID MAPPING PROGRAMS

4.1 RHMAP

RHMAP (Boehnke *et al.*, 1991) is a set of three FORTRAN 77 programs which comprise a two-point, minimum breaks and maximum likelihood analysis of RH mapping data. The package is freely available at: http//www.sph.umich.edu/group/statgen/software/. RHMAP programs are available in a form appropriate for IBM-compatible microcomputers running DOS. On these machines the programs can be installed simply by copying the three executable files RH2PT.EXE, RHMINBRK.EXE and RHMAXLIK.EXE to the hard disk. RHMAP can also be installed on computers using a UNIX operating system.

The two-point analysis (RH2PT) program is ideal for an initial analysis of a RH dataset. This will indicate an initial likely marker order and identify groups of closely linked markers which can then be analysed further using minimum breaks and maximum likelihood methods. Both the minimum breaks (RHMINBRK) and maximum likelihood(RHMAXLIK) methods consider all loci simultaneously rather than restricting analysis to two or four at a time. The minimum breaks and maximum likelihood can be used separately as distinct approaches to identifying the best locus order. Alternatively, the minimum breaks method can be used to obtain a preliminary set of candidate orders which can then be further analysed using the maximum likelihood method.

RHMAP supports the following retention models: (1) equal retention, (2) left-end, (3) centromere, (4) general, (5) selected marker and (6) multiple panels. The first four models are discussed in Section 4.1.4. The last two models are additions to RHMAP 3.0 (Lunetta *et al.*, 1996) and are not included in this manuscript. For haploid data, models 1–3 are Markovian in the sense that the likelihood of a locus is conditional on the previous locus. The general retention model is non-Markovian and is not supported for diploid data. For models 1–3, the likelihood for diploid data is computed using the theory of hidden Markov chains (Baum, 1972). RHMAP uses the EM algorithm (Dempster *et al.*, 1977) iteratively to maximize the likelihood; on each iteration the breakage and retention probabilities are updated. Algorithms for haploid data are described by Boehnke *et al.* (1991) and for diploid data by Lange *et al.* (1995).

4.1.1 Data

To demonstrate RH map generation using RHMAP, we will run a real data set (Figure 6.2) through the various analysis methods. This dataset contains the

	Markers										Markers							
	1	2	3	4	5	6	7	8		1	2	3	4	5	6	7	8
1	-	-	-	-	-	-	-	-	48	-	-	-	-	+	+	+	+
2	-	+	-	-	+	+	+	+	49	+	+	-	-	-	-	+	-
3	-	-	-	-	?	-	+	+	50	-	-	-	-	-	-	-	-
4	-	+	?	-	-	-	-	-	51	-	-	-	-	-	-	-	+
5	-	+	-	+	+	-	-	-	52	+	+	-	-	?	-	-	+
6	-	-	-	-	-	-	-	-	53	-	-	+	+	-	-	-	-
7	-	+	+	+	-	-	-	-	54	-	-	-	-	-	-	-	-
8	+	+	+	+	-	-	-	+	55	-	-	+	-	-	+	-	-
9	-	+	+	+	-	-	-	-	56	-	+	+	-	-	+	-	-
10	-	+	-	-	-	-	-	-	57	-	-	-	-	-	-	+	-
11	-	-	-	-	-	-	-	-	58	-	+	-	-	-	+	+	+
12	-	-	-	-	-	-	-	-	59	-	-	-	+	-	-	-	-
13	-	-	-	+	-	-	-	-	60	+	-	?	-	-	+	-	-
14	-	-	-	-	-	-	-	+	61	-	-	-	-	-	-	-	-
15	+	+	+	-	-	-	-	-	62	-	-	-	-	-	-	-	-
16	+	-	-	+	+	+	-	-	63	-	-	-	-	-	-	-	-
17	-	-	-	+	-	-	-	+	64	-	-	-	-	-	-	-	-
18	+	+	-	-	+	-	-	+	65	+	-	+	-	+	+	+	+
19	-	+	-	-	-	-	-	-	66	-	+	-	-	-	-	+	-
20	-	-	-	-	-	-	-	+	67	-	-	-	-	-	-	-	-
21	+	+	+	-	-	+	+	+	68	-	-	+	-	-	+	+	+
22	-	+	-	-	-	+	-	+	69	+	+	+	-	-	+	+	-
23	+	-	-	-	-	-	+	-	70	-	-	-	-	-	-	-	-
24	+	-	-	-	-	-	-	-	71	+	-	+	+	-	-	+	+
25	-	-	-	-	-	-	-	-	72	-	-	-	-	-	-	-	-
26	+	+	-	-	-	-	-	-	73	-	-	+	-	-	-	-	-
27	+	+	+	+	+	+	+	+	74	+	-	-	-	+	-	-	-
28	-	-	-	+	+	+	-	+	75	-	+	+	-	-	-	-	-
29	-	-	-	-	-	+	+	-	76	-	-	-	-	-	-	-	-
30	-	-	-	-	-	?	+	+	77	+	+	+	-	+	+	+	+
31	-	-	-	-	-	-	?	+	78	+	+	-	-	-	-	-	-
32	-	-	-	-	-	-	-	-	79	-	-	-	-	-	-	-	-
33	-	-	-	-	-	-	-	-	80	-	+	-	-	+	-	-	-
34	+	+	+	+	-	-	+	+	81	-	-	-	-	-	?	-	-
35	-	+	-	-	-	-	-	-	82	-	-	-	-	+	-	-	-
36	-	-	-	-	-	-	-	-	83	+	-	-	-	-	-	-	-
37	-	-	+	-	-	-	-	-	84	-	-	-	-	-	-	-	-
38	+	-	+	+	+	+	+	-	85	-	-	+	-	-	-	-	-
39	+	-	-	+	-	+	-	-	86	-	-	-	-	-	+	-	-
40	+	+	+	-	-	-	-	-	87	-	-	-	-	+	+	+	+
41	-	+	+	+	+	+	-	-	88	-	-	-	-	-	-	-	-
42	-	-	-	-	-	-	-	-	89	-	-	-	-	-	-	-	-
43	-	+	-	-	-	-	-	-	90	-	-	-	-	-	-	-	-
44	-	-	+	-	+	-	-	+	91	-	-	-	-	-	-	-	-
45	-	-	+	-	-	-	-	-	92	+	+	+	-	-	+	+	+
46	-	-	-	-	-	-	-	-	93	-	-	-	+	-	+	+	+
47	-	+	-	-	-	-	-	-	94	+	+	+	+	-	-	-	-

Figure 6.2 *Retention patterns for markers 1–8 are shown for each of the 94 hybrids. To demonstrate the variation in fragment length, the markers are represented in the order in which they are localized on the genetic map such that marker 1 is closest to the centromere and marker 8 furthest from it. The markers span an approximately 75–100 Mbp region (50 cM).*

results of eight chromosome 2 markers screened on 94 hybrids from the 'T31' mouse WG-RH panel (McCarthy *et al.*, 1997 (available from Research Genetics, 2130 Memorial Parkway SW, Huntsville, AL 35801, USA)). This panel was generated using 3000 rads of X-rays from an X-ray tube source. These markers were previously placed on the mouse genetic map, and spanned a 50 cM region. In mouse 1 cM is approximately equivalent to 1.5–2 Mb, so on average the eight markers are situated 9.5–12.5 Mb apart. Given the high resolution of the RH panel, these distances are near or at the outer limits of high confidence resolution for RH mapping. We have chosen this dataset to demonstrate the performance of the mapping packages under 'real life' conditions, which rarely generate a perfect map first time.

4.1.2 Two-Point Analysis

RH2PT provides two-point lod scores for linkage of all marker pairs, linkage groups, estimates of locus-specific retention probabilities and pairwise breakage probabilities. This allows a quick estimation of the data quality and indicates which retention models should be used for maximum likelihood multipoint analysis.

In this analysis the following assumptions are made to estimate breakage probabilities, distances and calculation of maximum lod scores. These are that breakage occurs at random along the chromosome; different chromosomal fragments are retained independently in hybrids; and retention probabilities for all chromosomal fragments are equal.

For detailed information on formating input files and running all programs in the RHMAP package, consult the manual distributed with the software. An edited version of an input file for a dataset containing eight markers typed on 94 hybrids is given in Figure 6.3.

```
    8   94    1    2
312 149 297f323 182 300f44f 274
(A8,1X,8(A1),T9,I1)
+-?
312 149 297f323 182 300f44f 274
        94  ++++----
        93  ---+-+++
        92  +++--+++
    .       ........
    .       ........
    .       ........
         3  ----?-++
         2  -+--++++
         1  --------
```

Figure 6.3 *Abbreviated input file for RHMAP two-point analysis.*

Line 1 of the input file contains the number of loci, number of hybrids, the output option and the ploidy for these data. Line 2 contains a list of all locus names in the data set. Line 3 contains a FORTRAN statement for reading the hybrid names and retention status data. Line 4 contains the status characters representing (1) locus typed and present, (2) locus typed and absent, and (3) locus not typed. In this example these characters are +, − and ?. Line 5 contains a list of all locus names in the order in which you would like them to appear in the output tables. Lines 6 to the end of the file contain the hybrid records specifying the hybrid name, and retention information for each hybrid.

To run the program type 'rh2pt'. You are then prompted for the input file name and output file name.

The output file from RH2PT consists of seven tables. The first three tables reiterate the input information. The fourth table prints the locus retention probabilities. This contains a list of the loci, the number and proportion of hybrids typed for each locus, and the number and proportion of typed hybrids that retain each locus. This output is shown in Figure 6.4, and gives an average retention of 25.4% for the eight markers on the panel.

The fifth output table prints the conditional coretention probabilities for each locus pair. These are the probability that the first locus is retained given that the second locus is (is not) retained, and the probability that the second locus is retained given that the first locus is (is not) retained.

This is followed by a table containing the maximum lod scores, breakage probability and distance estimates (in Rays) for each locus pair (Figure 6.5). The distance between the loci is given in Rays. As previously mentioned, equal retention is assumed for all fragments, so these figures are likely to differ from those obtained after more stringent multipoint analysis.

The final output table prints linkage groups based on the results of table 6. Linkage groups contain loci for which there is clear evidence of pairwise linkage.

LOCUS RETENTION PROBABILITIES

				P(RETAINED)	
LOCUS	TYPED	P(TYPED)	RETAINED	OVERALL	HAPLOID
312	94	1.000	26	0.277	0.149
149	94	1.000	31	0.330	0.181
297f	92	0.979	26	0.283	0.153
323	94	1.000	19	0.202	0.107
182	92	0.979	16	0.174	0.091
300f	92	0.979	23	0.250	0.134
44f	93	0.989	22	0.237	0.126
274	94	1.000	26	0.277	0.149
TOTAL	745	0.991	189	0.254	0.136

Figure 6.4 *Locus retention probability output from RHMAP two-point analysis.*

MAXIMUM LOD SCORES AND BREAKAGE PROBABILITY AND
DISTANCE ESTIMATES

LOCUS1	LOCUS2	BOTH TYPED	--	-+	+-	++	P(BR)	DIST	LOD SCORE
312	149	94	52	16	11	15	0.660	1.078	3.11
312	297f	92	55	12	11	14	0.605	0.928	3.74
312	323	94	58	10	17	9	0.777	1.500	1.47
312	182	92	58	9	18	7	0.839	1.825	0.88
312	300f	92	53	13	16	10	0.794	1.580	1.36
312	44f	93	56	11	15	11	0.715	1.255	2.17
312	274	94	52	16	16	10	0.840	1.831	1.01
149	297f	92	50	12	16	14	0.700	1.203	2.51
149	323	94	52	11	23	8	0.921	2.535	0.43
149	182	92	53	9	23	7	0.922	2.556	0.39
149	300f	92	48	13	21	10	0.882	2.138	0.71
149	44f	93	50	12	21	10	0.861	1.971	0.86
149	274	94	48	15	20	11	0.871	2.049	0.82
297f	323	92	58	8	15	11	0.661	1.081	2.75
297f	182	90	54	10	20	6	0.927	2.620	0.33
297f	300f	90	53	11	15	11	0.723	1.284	2.04
297f	44f	91	53	12	16	10	0.779	1.510	1.47
297f	274	92	50	16	16	10	0.847	1.878	0.95
323	182	92	63	10	13	6	0.803	1.625	1.06
323	300f	92	57	16	12	7	0.856	1.937	0.77
323	44f	93	57	17	14	5	0.968	3.441	0.14
323	274	94	56	19	12	7	0.899	2.298	0.52
182	300f	90	61	13	6	10	0.607	0.935	3.25
182	44f	91	61	14	9	7	0.770	1.471	1.36
182	274	92	61	15	7	9	0.690	1.170	2.26
300f	44f	91	61	7	9	14	0.462	0.621	5.98
300f	274	92	57	12	10	13	0.602	0.921	3.68
44f	274	93	61	10	7	15	0.466	0.627	6.15

Figure 6.5 *Output from RHAMP two-point analysis: linkage statistics.*

This is estimated for maximum lod scores greater than 2.00, 3.00 and 4.00
respectively. The test dataset formed a single linkage group supported by lod
scores ⩾ 2.00. Clearly, marker 323 is not linked to any other markers at lod
scores ⩾ 3.00, and divides the marker set into three linkage groups under these
criteria. Only the final three markers remain in a single linkage group supported
by lod scores ⩾ 4.00 (Figure 6.6).

Any locus pairs that do not contain any obligate breaks (totally linked loci)

```
LINKAGE GROUPS

LOD SCORE CRITERION:    2.00
LINKAGE GROUP   1:
312    149    297f   323    182    300f  44f    274

LOD SCORE CRITERION:    3.00
LINKAGE GROUP   1:
312    149    297f
LINKAGE GROUP   2:
323
LINKAGE GROUP   3:
182    300f   44f    274

LOD SCORE CRITERION:    4.00
LINKAGE GROUP   1:
312
LINKAGE GROUP   2:
149
LINKAGE GROUP   3:
297f
LINKAGE GROUP   4:
323
LINKAGE GROUP   5:
182
LINKAGE GROUP   6:
300f   44f    274
```

Figure 6.6   *RHMAP two-point linkage groups.*

are printed at the end of the output file along with their coretention pattern. If these loci contain missing data it should be decided at this point whether or not to retype the loci before carrying out further analysis of the dataset. In order to assemble a high confidence map, markers should form a single linkage group under two-point lod score criteria of at least > 4.000.

4.1.3   Minimum Breaks Analysis

RHMINBRK orders loci by minimizing the number of obligate breaks between loci, analogous to genetic mapping by minimizing recombinants. Counting the number of obligate breaks for every possible order is feasible only for relatively small numbers of loci, since for N markers there are $N!/2$ possible solutions. As N becomes large, the calculation becomes impractical; for example, if $N = 14$, the number of locus orders is more than 43 billion.

RHMINBRK contains four ordering strategies for generating locus orders.

1.   List of user-specified locus orders, where each candidate locus order specified in the input file is ranked based on minimizing obligate breaks.

2. Stepwise locus ordering, where locus orders are generated by adding one locus at a time, and only those partial orders that are within K breaks of the best partial order are retained for consideration. Increasing values for K can improve chances of achieving the best order, while also increasing the numbers of calculations required. When partial orders are eliminated, all orders descending from these partial orders are also eliminated. Consequently fewer partial orders are considered than when using the branch and bound approach, at the risk of missing the overall best order. Starting with an anchor map of well placed loci considerably improves the efficiency of the stepwise approach.

3. Simulated annealing starts with an n-locus order, and initially randomly inverts blocks of loci, and assesses the number of obligate breaks at each inversion step. A list of best encountered orders is kept during the process.

4. Branch and bound is similar to stepwise ordering. The branch and bound analysis is primed using a greedy algorithm to generate an initial reasonable candidate order. To guarantee the best solution, all permutations of markers must be generated. To do this systematically, a search tree is constructed where each branch is a permutation; for example, given markers {a,b,c,d}, the branches are {abcd, abdc, acbd, acdb, ...}. A partially created branch is referred to as a node. If the current node has more obligate breaks than the best solution so far, all branches below the node need not be generated as they will obviously not produce a permutation with a lower score (e.g. if the partial solution 'ab' has more obligate breaks then the best solution, any branch starting with 'ab', can be eliminated automatically). Branch and bound is the only ordering strategy that can guarantee that the best locus order is identified. Its limitation is that it requires a lot of computation for large numbers of loci. Although not all the $N!/2$ possible solutions need to be generated, the number of permutations generated actually depends on the quality of the input and generally remains intractable.

The input file for RHMINBRK can contain information for multiple analyses of each of several datasets. Figure 6.7 contains an example of an input file in which three separate analyses are specified. The input file consists of two parts, the dataset and the problem control record. The dataset contains few differences with the input file for RH2PT.

The problem control record contains information required for each of the analyses to be performed on the dataset. These include the number of loci in the problem, the ordering option (e.g. stepwise ordering), the upper limit on the number of locus orders to print, and locus names for all loci in the problem set. At this stage there is also an option to exclude hybrid data containing ambiguous data (marked: ?). Also, for each analysis problem one of four sets of additional information is required, depending on which ordering option is chosen. This additional information includes options to use a machine-generated candidate

```
3   8  94   1
312 149 297f323 182 300f44f 274
(A8,1X,8(A1),T9,I1)
+-?
        94 ++++----
        93 ---+-+++
        92 +++--+++
           ..........
           ..........
           ..........
         3 ----?-++
         2 -+--++++
         1 --------
      8   2   1   0   1          ‹ Problem 1: Stepwise ordering
    312 149 297f323 182 300f44f 274
      0   0  10    5

      8   2   1   0   1          ‹ Problem 2: Stepwise ordering with marker 1
    312 149 297f323 182 300f44f 274    forced to be placed at the beginning of the map
      0   0  10    3  -1
    312

      8   4   1   0   1          ‹ Problem 3: Branch and bound ordering
    312 149 297f323 182 300f44f 274
      0   0  10    5
    312 149 297f323 182 300f44f 274
```

Figure 6.7   *Abbreviated input file for RHMAP minimum breaks analysis.*

order or a user-specified order, as well as an option to force a subset of the markers into user-specified positions in the final map order.

The RHMINBRK program is run by typing 'rhminbrk'; you are then prompted for input and output file names.

The RHMINBRK output file contains a list of locus orders with minimum obligate breaks. Each order is ranked beginning with the best order. The number of obligate breaks is listed for each order, followed by the locus order. The locus number is printed under each locus name.

Problem 1 specified stepwise ordering, saving orders within 10 obligate breaks of the best order, and printing all orders within five obligate breaks of the best order. The results of this analysis are displayed in Figure 6.8. There are four possible orders within five obligate breaks of the best scoring order. We know from the genetic map locations of these markers that the correct order is: 1, 2, 3, 4, 5, 6, 7, 8. Markers 1 and 2 are in the wrong order because of the large physical distance between the first three markers.

In datasets like this, where the correct order is known for some or all of the markers, it can be informative to force one or more markers into specific positions in the order. When the first marker (312) is forced to be placed at the beginning of each order in problem 2, the optimal order is identical with genetic map order (Figure 6.9). The second ranked order places marker 2 (149) at the end of the map. This raises a potential problem with minimum breaks analysis.

```
ORDERING OPTION:                    STEPWISE
MAXIMUM BREAK DIFFERENCE TO SAVE ORDER:   10
MAXIMUM BREAK DIFFERENCE TO PRINT ORDER:   5
ADDING ORDER:              MACHINE-GENERATED
CANDIDATE ORDER:           MACHINE-GENERATED
ORDER FOR ADDING LOCI:     312   149   323   274   182   297f 300f 44f


LIST OF BEST MINIMUM OBLIGATE BREAK LOCUS ORDERS
RANK   BREAKS   LOCUS ORDER

  1      151     149   312   297f 323   182   300f 44f   274
                  2     1     3    4     5     6    7     8

  2      155     149   312   297f 323   182   274   44f   300f
                  2     1     3    4     5     8     7     6

  3      156     149   312   297f 323   182   300f 274   44f
                  2     1     3    4     5     6    8     7

  4      156     312   149   297f 323   182   300f 44f   274
                  1     2     3    4     5     6    7     8
```

Figure 6.8   *Abbreviated output from RHMAP minimum breaks analysis.*

Figure 6.10 gives a graphical representation of the retention rate for each marker as presented in the RH2PT output table. Both markers 1 and 3 have retention rates of approximately 0.28, with marker 2 having a retention rate of 0.33. The large physical distance between markers 1, 2 and 3 places them at the outer limits of high confidence resolution using this hybrid panel. Looking at the scored data in Figure 6.2, there are 12 positives for marker 2 on hybrid DNA fragments which do not extend to markers 1 and 3. This 18% increase in retention

```
ORDERING OPTION:                    STEPWISE
MAXIMUM BREAK DIFFERENCE TO SAVE ORDER:   10
MAXIMUM BREAK DIFFERENCE TO PRINT ORDER:   3
ADDING ORDER:              MACHINE-GENERATED
CANDIDATE ORDER:           MACHINE-GENERATED
ORDER FOR FORCED LOCI:     312
FIRST LOCUS 312   IN THIS ORDER IS FORCED TO BE AT AN END OF THE MAP.
ORDER FOR ADDING LOCI:     312   297f 274   300f 44f   182   323   149


LIST OF BEST MINIMUM OBLIGATE BREAK LOCUS ORDERS
RANK   BREAKS   LOCUS ORDER

  1      156     312   149   297f 323   182   300f 44f   274
                  1     2     3    4     5     6    7     8

  2      159     312   297f 323   182   300f 44f   274   149
                  1     3     4    5     6     7    8     2
```

Figure 6.9   *Abbreviated output from RHMAP minimum breaks analysis.*

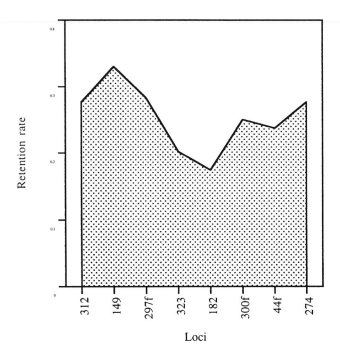

Retention rate for loci on T31 radiation hybrid panel

☷ Estimated retention rate of the hybrid panel accross
the 75-100 Mb region spanned by the 8 loci.

Figure 6.10 *Variation in retention rate for the eight markers spanning a
75–100 Mbp region of mouse chromosome 2. The markers are listed in the order
in which they are localized on the genetic map.*

generates a high number of obligate breaks between marker 2 and its flanking
markers. The order generated using minimum breaks analysis is therefore more
likely to place marker 2 at one end of the marker order. In datasets where the
distance between markers is large for the outer resolution of the hybrid panel,
minimizing breaks will not necessarily give the true marker order. In these situ-
ations the marker density should be increased until high-confidence linkage is
obtained. Figure 6.9 contains the output after stepwise analysis of the dataset,
specifying that marker 1 must be forced to the beginning of the order.

Problem 3 in this analysis specified the branch and bound ordering option.
Orders are printed that are within a maximum of five obligate breaks more than
the optimal order. The four best orders are identical with those obtained using
stepwise analysis (Figure 6.11).

```
ORDERING OPTION:                     BRANCH
MAXIMUM BREAK DIFFERENCE TO SAVE ORDER:   10
MAXIMUM BREAK DIFFERENCE TO PRINT ORDER:   5
ADDING ORDER:            MACHINE-GENERATED
CANDIDATE ORDER:         MACHINE-GENERATED
CANDIDATE LOCUS ORDER:   274  44f  300f 182  323   297f 312  149
ORDER FOR ADDING LOCI:   312  149  323  274  182   297f 300f 44f


LIST OF BEST MINIMUM OBLIGATE BREAK LOCUS ORDERS
RANK   BREAKS   LOCUS ORDER

  1     151     149  312  297f 323  182  300f 44f  274
                 2    1    3    4    5    6    7    8

  2     155     149  312  297f 323  182  274  44f  300f
                 2    1    3    4    5    8    7    6

  3     156     149  312  297f 323  182  300f 274  44f
                 2    1    3    4    5    6    8    7

  4     156     312  149  297f 323  182  300f 44f  274
                 1    2    3    4    5    6    7    8
```

Figure 6.11   *Abbreviated output from RHMAP minimum breaks analysis.*

If information is requested on influential hybrids (INFOPT = 1), the following tables are printed:

1. Retention data permuted in the best locus order for the problem.
2. Numbers of obligate breaks per hybrid. This table presents the numbers of obligate breaks per hybrid. If large numbers of hybrids require multiple breaks, it is worth re-examining the typing of these hybrids for errors.
3. Influential hybrids for the most likely orders. This table lists for all orders other than the best order, the hybrids that require a different number of obligate breaks compared with the best order. Any hybrids solely responsible for the relative ordering of loci are worth retyping for possible errors.

4.1.4   Maximum Likelihood Analysis

RHMAXLIK carries out a maximum likelihood analysis for four different breakage and retention models. All of these models assume that X-ray breaks occur as a Poisson process. The retention models all make different assumptions about DNA fragment retention across the chromosome. These models can be selected based on the observed retention for each locus provided in the RH2PT output file. The models range in complexity, from assuming that all retention probabilities are equal to assuming that all retention probabilities may differ. It is important to select appropriate retention models to avoid unnecessary distortion of distance estimates. For example, a significant difference in retention between adjacent loci will result in a large number of obligate breaks. Analysis of these data using an equal retention model would result in an expansion of the distance estimates between these loci, compared with a centromeric or general retention model. The following is a brief description of these models:

1. The equal retention model assumes that all fragments have the same retention probability.
2. The centromeric model allows for a higher retention probability for fragments containing one endpoint of the map. A higher than average retention rate is frequently observed for fragments containing centromeric, and to a lesser degree telomeric, DNA.
3. The left-endpoint retention model allows the fragment retention probability to depend on the left-most locus present in the fragment.
4. The general retention model (Cox *et al.*, 1990) allows all retention probabilities to differ. It has been proposed that this model generates more accurate map distances for all data where the retention probability is not equal; that is to say that for datasets showing centromeric or telomeric retention, the general model generates more accurate distance estimates than both the centromeric and left-end retention models (H. Jones, personnal communication). The general retention model is supported only for haploid data.

The ordering strategies for RHMAXLIK are the same as those available for RHMINBRK:

1. List of user-specified locus orders.
2. Stepwise locus ordering.
3. Simulated annealing.
4. Branch and bound.

The first portion of the RHMAXLIK input file is essentially the same as for RHMINBRK. The hybrid retention data are followed by the problem control records. As in RHMINBRK, a number of different analyses can be performed on the dataset. A retention model must be specified for each analysis problem, as well as an ordering option. In this analysis equal retention and general retention models will be performed using both stepwise ordering and branch and bound ordering. RHMAXLIK analysis is run by typing 'rhmaxlik'.

You are then prompted for an input filename, an output filename and an iteration filename. The iteration file exists for the purpose of running the program and should be deleted after the analysis is complete.

Most tables in the output file are self-explanatory. Figure 6.12 contains an

```
NUMBER OF LOCI:                           8
NUMBER OF CHROMOSOMES (PLOIDY):           2
RETENTION MODEL:                 EQUAL
ORDERING OPTION:                 STEPWISE
GENETIC LOCI:    312  149  297f 323  182  300f 44f  274
MAXIMUM LOG10-L DIFFERENCE TO SAVE ORDER:    6.00000
MAXIMUM LOG10-L DIFFERENCE TO PRINT'ORDER:   4.00000
MINIMUM LOG10-L SUPPORT TO ADD LOCUS:        0.00000
ORDER FOR ADDING LOCI:   300f 44f  274  182  323  297f 149  312


MOST LIKELY LOCUS ORDERS


                 LOG10
           LIKE  LIKE
RANK  DIFF RATIO BRKS  LOCUS ORDER

  1 0.0000 1.0   151    149  312  297  323  182  300f 44f  274
                         2    1    3    4    5    6    7    8
 ..    ..   ..    ..    ..........................................
 ..    ..   ..    ..    ..........................................
 ..    ..   ..    ..    ..........................................
  6 1.0433 11.0  156    312  149  297f 323  182  300f 44f  274
                         1    2    3    4    5    6    7    8
 ..    ..   ..    ..    ..........................................
 ..    ..   ..    ..    ..........................................
 ..    ..   ..    ..    ..........................................


PARAMETER ESTIMATES FOR THE MOST LIKELY LOCUS ORDERS

RANK         LOCUS ORDER
 .    .      .    .     .    .     .    .     .    .     .
 .    .      .    .     .    .     .    .     .    .     .
 .    .      .    .     .    .     .    .     .    .     .
  6          312  149   297f 323    182   300f   44f    274
BRK            0.672 0.712 0.657 0.743 0.582 0.464 0.470
DIST           1.115 1.246 1.069 1.357 0.873 0.623 0.635
RETOBS  0.277 0.330 0.283 0.202 0.174 0.250 0.237 0.277
RETEST  0.138 0.138 0.138 0.138 0.138 0.138 0.138 0.138
RETPAR  0.138
TOTAL MAP LENGTH:   6.919
 .    .      .    .     .    .     .    .     .    .     .
 .    .      .    .     .    .     .    .     .    .     .
 .    .      .    .     .    .     .    .     .    .     .
```

Figure 6.12   *Abbreviated RHMAP maximum likelihood analysis output file.*

abbreviated example of the output for problem 1. Following a list of the most likely orders, ranked from most likely to least likely within the specified parameters, is a list of these orders along with breakage probability and distance estimates between consecutive markers, as well as the observed and estimated retention for each marker.

For all models except the general retention model, the retention probability is initially estimated (RETEST) as the sample proportion of retained loci among the appropriate typed loci (Boehnke *et al.*, 1991). For the general retention model, RETEST is calculated using the moment estimates suggested by Cox *et al.* (1990).

The distance estimates (DIST) given in Rays are generally presented as cRays on genomic maps. The actual retention rate (RETOBS) for each marker varies from 0.174 to 0.330 and does not fit the equal or centromeric retention models. The above analysis was performed using the equal retention model. To demonstrate the importance of choosing the correct model for each dataset, these data were reanalysed using the general retention model. As mentioned previously, the general retention model is supported only for haploid data. This is a diploid dataset, so it must be emphasized that analysing the data as haploid will not provide accurate distance estimates but may indicate more likely orders, given the retention pattern of this dataset. An abbreviated version of the output file containing the stepwise analysis tables under the general retention model is given in Figure 6.13.

The known correct order was ranked sixth using the equal retention model with a map distance of 6.919 R. The order was ranked second using the general retention model, which gave a map distance of 7.318 R. With the positioning of marker 2 at the beginning of the optimal order under the general retention model, it is clear that additional markers must be typed in at least the region spanned by the first three markers in order to generate a high confidence map.

Simulated annealing and branch and bound ordering of the same dataset using the equal retention model generated an identical list of orders to that obtained by stepwise analysis using the equal retention model.

RHMAP is most suitable for the analysis of smaller datasets, as indicated by the default maximum array dimensions for RHMAP:

RH2PT
maximum number of hybrids in a dataset	200
maximum number of loci in a dataset	60
maximum number of locus pairs in a dataset	1770

RHMINBRK
maximum number of hybrids in a dataset	200
maximum number of loci in a dataset	60
maximum number of locus orders to process	1000

RHMAXLIK
maximum number of hybrids in a dataset	100
maximum number of loci in a dataset	32
maximum number of locus orders to process	1000
maximum number of model parameters	64

These dimensions can be altered, but if they are surpassed the execution can become cumbersome.

```
NUMBER OF LOCI:                           8
NUMBER OF CHROMOSOMES (PLOIDY):           1  N.B. For comparison purposes.
RETENTION MODEL:                    GENERAL
ORDERING OPTION:                    STEPWISE
GENETIC LOCI:    312  149  297f 323  182  300f 44f  274

MOST LIKELY LOCUS ORDERS

         LOG10
         LIKE   LIKE
RANK     DIFF   RATIO BRKS  LOCUS ORDER

  1     0.0000  1.0   151    149  312  297f 323  182  300f 44f  274
                              2    1    3    4    5    6    7    8

  2     0.6724  4.7   156    312  149  297f 323  182  300f 44f  274
                              1    2    3    4    5    6    7    8
 ..      ..     ..    ..    ........................................
 ..      ..     ..    ..    ........................................
 ..      ..     ..    ..    ........................................


PARAMETER ESTIMATES FOR THE MOST LIKELY LOCUS ORDERS

RANK        LOCUS ORDER
  1        149     312  297f    323    182  300f    44f    274
BRK            0.763 0.605 0.564 0.700 0.575 0.467 0.692
DIST           1.438 0.929 0.831 1.203 0.854 0.630 1.176
RETOBS    0.330 0.277 0.283 0.202 0.174 0.250 0.237 0.277
RETEST    0.325 0.271 0.270 0.199 0.167 0.248 0.230 0.275
RETPAR    0.272
RETPAR    1.000 0.273
RETPAR    1.000 0.000 0.440
RETPAR    0.040 0.000 0.000 0.244
RETPAR    0.021 1.000 0.000 0.000 0.185
RETPAR    0.000 0.185 0.000 0.000 1.000 0.318
RETPAR    0.000 0.000 1.000 0.318 0.069 0.613 0.196
RETPAR    0.000 1.000 0.318 0.069 0.613 0.196 0.262 0.187
TOTAL MAP LENGTH:      7.062

  2        312     149  297f    323    182  300f    44f    274
BRK            0.679 0.732 0.612 0.715 0.598 0.571 0.597
DIST           1.136 1.316 0.946 1.254 0.911 0.845 0.909
RETOBS    0.277 0.330 0.283 0.202 0.174 0.250 0.237 0.277
RETEST    0.269 0.322 0.270 0.200 0.166 0.247 0.232 0.273
RETPAR    0.247
RETPAR    1.000 0.390
RETPAR    0.000 0.000 0.302
RETPAR    0.000 1.000 0.000 0.201
RETPAR    0.032 0.655 0.000 0.000 0.170
RETPAR    0.000 0.170 0.000 0.000 1.000 0.237
RETPAR    0.000 0.000 1.000 0.237 0.121 0.564 0.187
RETPAR    0.000 1.000 0.237 0.121 0.564 0.187 0.142 0.221
TOTAL MAP LENGTH:      7.318
```

Figure 6.13 *Abbreviated output from RHMAP maximum likelihood analysis using the general retention model.*

4.2   RHMAPPER

RHMAPPER is a recently developed RH mapping program (Stein *et al.*, 1995; Slonim *et al.*, 1996) designed to accommodate thousands of markers in large-scale map generation. It has been used to create a whole genome map at the Whitehead Institute/MIT Center for Genome Research (Data release II, http://www.genome.wi.mit.edu/ftp/distribution/human_STS_release/oct96/) (Hudson *et al.*, 1995) and dense long-range chromosome-specific maps at the Sanger Centre (http://www.sanger.ac.uk/humanmap/rhmap/). RHMAPPER is an interactive command-line driven program which provides various commands to query the data and test various hypotheses. It supports building a framework and then placing the remaining markers within the framework, resulting in a placement map.

The RHMAPPER interface is a Perl interpreter, version 5.003 or higher. The database is the Berkeley DB which allows arbitrary keyword/data pairs to be stored in files. Perl5 comes with the interface to the DB_File, but the DB library may need to be compiled and installed. The computationally intensive work is performed by a C program with a client–server interface to the Perl program which is invisible to the user. The salient software for RHMAPPER is easy to install; additional software can be installed to provide a graphical display. The software has been tested on a Sun Sparc10 running SunOS 4.3, Dec Alpha running OSF/1 2.0 and 3.0, and Intel Pentium CPU running Linux 1.2.13.

Using RHMAPPER commands, the initial framework map can be computed by finding well ordered triples (groups of three markers) and then ordering the triples using a graph theoretical approach. This creates a sparse framework that can be 'grown' by greedily trying to add additional markers. The remaining markers can then be tested individually to determine which ones can be binned within an interval of the framework. All evaluations of a given order are calculated by the C program, which uses a hidden Markov model to perform maximum likelihood calculations on multipoint maps, where the 'hidden' states can represent uncertainties in the data. The analysis makes the following assumptions: the retention of a marker depends only on the retention of the previous marker and the probability of a break between the two markers; the retention is taken to be constant for each hybrid, but different hybrids may have different retention rates; radiation-induced breaks occur randomly along a chromosome as a Poisson process and different fragments are retained independently in a given hybrid. An EM algorithm is used to find the most likely map for a given order. This is similar to the process used by Lander and Green (1991) to build genetic maps. For further information on the algorithms, see Slonim *et al.* (1996).

RHMAPPER is well documented and a description of the commands can be found in the manual that comes with the software (Stein *et al.*, 1995). The following is intended to give an overview of software usage and is not intended to be all inclusive of its capabilities. The data and parameters will be described, and two examples of RHMAPPER analysis are provided: the first example uses a dataset of 73 markers screened on a baboon panel and the second uses a dataset of eight chromosome 2 markers screened on the mouse T31 panel, as per the RHMAP description.

4.2.1   Data

The input is a file containing markers and their associated retention patterns. The retention pattern is a vector of $\{0,1,2\}$s representing negative, positive and unknown scores. The unknown data are either missing or ambiguous (e.g. differing scores from duplicate experiments). The first line of the file contains the keywords. The subsequent lines contain the values for each marker. A sample input is shown in Figure 6.14a, the first marker has keyword/value pairs of NAME = f8vwfp and RHVECTOR = 000101 ... 000. Additional site-specific keywords/value pairs can also be present and are stored with each marker in the database (see Section 4.2.3). A test data file is in the directory 'testdata' which is in the RHMAPPER tar file.

The RHMAPPER distribution tar file contains the directory structure shown in Figure 6.14b. By default, the data files go into the directory DATA. There is

a.

```
NAME RHVECTOR
f8vwfp     000101000100000201010011000010100000100000010000000010010001000100000000
igkvp3     0000000000001000000010111100010110000000000010000000101000100000000000100
atp6e      0010010001000000010110111000000000000100000000000000001100100101101000001 01
tuple1     0000000000010000100010011000010000001000000100000010100100010101010000100
comt       0001000000010000000010010100010000001000000000000000001001100001010001100
etc.
```

b.

```
RHMAPPER-1.1/rhmap:
DATA/          commands/        RHMAPPER.conf     testdata/
Makefile       local_hacks/     RHMAPPER.template  util/
bin/           rhmapper*        server/
```

c.

```
DATA/baboon:
Keywords     Marker_info  Marker_names  Pairs    groups/  maps/

DATA/baboon/groups
link     fw

DATA/baboon/maps
fw.map
```

Figure 6.14 *(a) An example input file. (b) The RHMAPPER directory structure. (c) A project directory which contains DB_Files and text files.*

a subdirectory for each project. After the analysis of example 1 (see Section 4.2.3), the directory DATA will contain the files shown in Figure 6.14c. The files directly under the baboon directory are the DB_Files which contain the keyword/value pairs from the input file. The files in the 'groups' directory are text files in which the format is simply a group name followed by one or more markers. There can be multiple groups in a file. Many commands take group names as input to save the user from having continually to retype a set of marker names. The most important group is framework markers. The 'maps' directory contains one or more text files which represent placement maps.

4.2.2   Parameters

To execute RHMAPPER, go to directory RHMAPPER-1.1/rhmap and type 'rhmapper'. It immediately reads the file 'rhmapper.conf' which contains most of the parameters and default settings. The parameters can be changed by editing the file or interactively by using the **set** command. RHMAPPER uses a constant retention frequency (default 0.4). The HHM algorithm uses parameters for the probability of false-positive, false-negative and missing data. The framework and placement commands have parameters as described below (the upper case words are parameters). The ploidy can be changed from diploid to haploid by executing the command **haploid**.

4.2.3   Example 1: Baboon Panel

The data for the first example covers the short arm of baboon chromosome 13 (D. Spillett and P. Hayes, unpublished data). It is a diploid panel with a 20% retention frequency and was generated using 3000 rads. The data file contains 73 markers and 73 hybrids. The set of commands shown in Figures 6.15–6.20 build an initial placement map. The only parameter changed is the retention frequency, which is set to 0.2. This example shows that initial results can be computed easily. It is not intended to represent the final map.

In the following baboon analysis, the paragraphs starting with numbers 1–4 explain the commands and output shown in Figure 6.15. Similarly, paragraphs 5–9, 8–9, 10–13, 14 and 15–16 explain figures 6.16, 6.17, 6.18, 6.19 and 6.20 respectively.

1> The first command to RHMAPPER must be to set the project name. Setting the project to 'baboon' will cause subsequent commands to create files in 'DATA/baboon' or access existing files in this directory.

2> The **load_data** command initializes the DB_Files in the baboon directory with the data in the file 'testdata/baboon'. The '1' tells the program to compare

```
1> set project baboon
PROJECT = baboon

2> load_data testdata/baboon.dat 1
./DATA/baboon doesn't exist. Creating directory.
Loaded: 73
2619 pairs loaded.
{...}

3> link_groups
Found 1 linkage groups at LOD=3.0.
Group 1: size 73
{73 markers listed}
UNLINKED: 0
Creating groups named LINK_1 to LINK_1

4> saveg link
```

Figure 6.15 *An example RHMAPPER session. Beside each '>' is a command and following each command is zero or more parameters. Some of the output is verbose, in which case it is replaced by a short description in brackets.*

```
5> find_triples triples LINK_1
processing pairs ...
{output every 5000 triples}

Finally:
The best triple in database is:
wi-379 wi-183 d4f      -37.673197
wi-379 d4f wi-183      -41.863890
d4f wi-379 wi-183      -41.925087
with a lod score of 4.190693

6> assemble_framework1 triples
Loading triples...

There are 146 triples ordered at lod 3.0 and with no gaps larger than 30 cR.
Now searching for frameworks...

{listing of some paths found}

Overall longest path contains 5 markers.
Longest Path: ppar cryba4 az190we5 gnaz crkl3

7> defg fw ppar cryba4 az190we5 gnaz crkl3
```

Figure 6.16 *The RHMAPPER commands for building an initial framework.*

```
8> evaluate fw
NAME        BREAK FREQ    cR
ppar           0.228      25.9
cryba4         0.178      19.6
az190we5       0.172      18.9
gnaz           0.219      24.7
crkl3          0.000       0.0


LIKELIHOOD = -57.657005
MAP LENGTH = 89.07


9> ripple 3 fw
{a list of orders and scores}
GRAND BEST ORDER: ppar cryba4 az190we5 gnaz crkl3

NEXT BEST ORDER: cryba4 ppar az190we5 gnaz crkl3
LOD vs nextbest = 4.618106
```

Figure 6.17   *The RHMAPPER commands for verifying the integrity of a framework.*

all pairs of markers and to store the pairs that have a lod score above TWO_POINT_CUTOFF (lod 3.0 by default) in a file called 'Pairs'. This can take some time because it is executing $N(N - 1)/2$ comparisons, but then many subsequent commands run significantly faster as the pairwise lod scores can be accessed quickly from this file.

Commands 3–12 create the framework map. Creating the framework is the most important and time-consuming step. The initial framework can be seeded by a known order such as the genetic map; to do this, a group must be defined, as in command 7, containing the markers that have been localized previously. Various commands (e.g. see 5, 6, 8 and 9) should be run to verify this marker order and, if necessary, the group should be altered appropriately to provide the best framework. An alternative to seeding the initial framework with an ordered set of markers is to find linkage groups, make frameworks for each group and then merge them. We use the second approach in this example.

3> The link_groups command finds all sets of transitively linked markers. For each set, a group called 'LINK_N' is created where N is the group number. All markers in this example were found to be in one linkage group.

4> The saveq command saves the current set of groups in a file named 'link' On the next invocation of RHMAPPER, the file can be loaded back into memory by using the load_group command.

5> The find_triples command finds the well ordered triples from the markers in group 'LINK_1'. A well ordered triple is one in which one of the three possible orders is better than the other two by a lod score greater than FRAMETHRESH (default 3.0). The results are written to the file named 'triples'.

```
10> grow_framework fw LINK_1
{a bunch of trials similar to the following}
** yesp **
  Stepwise...
     ...failed with lod 1.21151400000001
** iglv2 **
  Stepwise...
  Rippling pairs...
  Rippling triples...
  SUCCEEDED: 031xc11 ppar cryba4 az190we5 gnaz iglv2 crkl3 comt wi-346
  LOD: 3.54939299999999
** wi-274 **
  Stepwise...
     ...failed with lod 0.338117999999994
{etc}
TRIED: 73
ADDED: 4
FINAL ORDER: 031xc11 ppar cryba4 az190we5 gnaz iglv2 crkl3 comt wi-346
Updating fw...

11> grow_framework fw LINK_1
TRIED: 73
ADDED: 0
FINAL ORDER: 031xc11 ppar cryba4 az190we5 gnaz iglv2 crkl3 comt wi-346
Updating fw...

12> evaluate fw
```

NAME	BREAK FREQ	cR
031xc11	0.350	43.1
ppar	0.232	26.4
cryba4	0.172	18.9
az190we5	0.165	18.0
gnaz	0.067	6.9
iglv2	0.159	17.3
crkl3	0.155	16.8
comt	0.367	45.7
wi-346	0.000	0.0

```
LIKELIHOOD = -96.278231
MAP LENGTH = 193.21

13> saveg fw
```

Figure 6.18   *The RHMAPPER commands for adding markers to a framework.*

6> The `assemble_framework1` command assembles the triples into the longest path that obeys the framework rules. A *path* is a term from graph theory and can be viewed as an alternative order in this context. The framework rules require a lod score greater than TRIPLE_LOD (default 3.0) and a distance less

```
14> create_placement_map fw.map LINK_1
CURRENT_GROUP = fw
Creating a new map named fw.map
{....}
Placing d22s268
d22s268 isn't linked to the framework at a cutoff of PLACEMENT_LINKAGE=5.0.
Placing d4f
PLACEMENTS FOR d4f:
```

LOD	FRAMEWORK	DISTANCE	NEW INTERVAL	OLD INTERVAL
0.000	ppar	14.5	26.4	26.4
0.704	cryba4	9.1	18.9	18.9
1.606	wi-346	95.5	95.5	0.0
1.822	031xc11	-98.9	98.9	0.0

```
Adding placement for d4f
Adding placement for d4f
d4f places too far off the end of the framework.
Discarding...
Placing bs320ve9
PLACEMENTS FOR bs320ve9:
```

LOD	FRAMEWORK	DISTANCE	NEW INTERVAL	OLD INTERVAL
0.000	cryba4	4.3	18.8	18.9
1.772	ppar	23.8	26.4	26.4

```
Adding placement for bs320ve9
Adding placement for bs320ve9
73 attempted. 16 placed.
```

Figure 6.19　*The RHMAPPER commands for creating a placement map.*

than TRIPLE_DIST (default 30 cR) between adjacent markers. It should be noted that this command goes into a endless loop quite easily. If the same message continues to print repeatedly, terminate the process by typing control-C, change the order of the triples in the triples file and try again.

7> The `defg` command makes a group named 'fw' of the list of markers. This group represents the ordered framework markers calculated in the previous command.

8–9> At this point, various tests should be run on the framework to ensure its absolute integrity. `Evaluate` calculates distances and likelihood for an ordered list of markers. The `ripple` commands slides a window across the markers and evaluates all permutations of markers within the window. In this case, the window size is 3 markers. As the next best order has a lod of 4.6, this order is over 10 000-fold more likely than the second best order.

10–11> The `grow_framework` command adds markers to the framework group. To be added, a marker must have a unique position greater or equal to the FRAMETHRESH (default 3.0) and it must pass the ripple test of window sizes 2 and 3. The markers in 'LINK_1' that are already in the

```
15>print_map
{Vectors are truncated on the right}
031xc11   29.60  F       00101100100010001000011000000100000000000000000000
164th8    13.48  P0.77   00000000100000001100111010000100000200000001000000000
ppar      11.77  F       20000000100000001000111010000100000010000000000000000
fibb       2.74  P0.62   00000000000000001100101010000100000000000000000000
d4f        0.00  P0.70   00000000000000001100100010010000001010000000000000000
lif        0.00  P0.47   00000000000000001100101010000000002010000000000000000
wi-274    11.89  P0.68   00000000000000001110101010000000101010000000000000000
cryba4     4.33  F       00000000000000001100110000000000000001000000000000000
bs320ve9   3.92  P1.77   00000000000000001100110000000000000000000000000000000
211yf10    0.00  P>3.00  00001100000010001100110000000000000000000001000000000
az298yb5   0.00  P1.82   00001100000010001100110000000000000000000000000000000
d22s351    0.00  P>3.00  00001100000010001100100000000000000001001000000000000
wi-125     0.00  P1.80   00001100000000001100100000000000000001000000000000000
wi-183     0.33  P1.60   00001100000000001100100000010000001010200000000000000
yesp       0.43  P2.10   00001100000010001100100200000000000001000000000000000
234vh2p    0.00  P>3.00  00000000000000001000000000000000000000000000000020000
merlin     9.88  P1.67   00000000000000001110100010000000001010000000000000000
az190we5  18.03  F       00000000000000001100100100000000000000000000000000000
gnaz       6.94  F       00010000000100001100100100000100000000000000002000000
iglv2      5.04  F       00010000000102001100100100000100001000000000000000000
iglv      12.28  P2.72   10010000000100001100100100000100001010000000000000000
crkl3      7.99  F       00010000000100000000100100000100021010000000000000000
hcf2       8.85  P>3.00  10010000000100000000100100000100000010000000000000000
comt      45.73  F       00010000000100000000100101000100000010000000000000000
wi-346     0.00  F       00010000000100001000100100001100000011001000000101000

16>save_map fw.map

17>quit
```

Figure 6.20 *The RHMAPPER commands for printing and saving a map.*

framework will be ignored. This command should be executed until no more markers are added.

12–13> The `evaluate` command is run on the final framework map to verify its integrity. Then the framework is saved in the file named 'fw'.

14> The `create_placement_map` command tries to place each marker in the input list into the framework without regard to other previously placed markers. All possible placement intervals are found for a marker, and the distance within each interval is computed. A marker is rejected if it fails the placement rules, which are: a marker must have a lod score greater than PLACEMENT_LINKAGE (default 5.0) with at least one framework marker; and a marker must be less than PLACEMENT_TOO_FAR cR (default 15.0) from the end of the framework.

15–16> The `print_map` command outputs the map. Framework markers are indicated by an 'F' in the third column. Placement markers are indicated by a 'P' in the third column followed by the lod score with the next best placement unless it is greater than 3.0. The second column contains distances in cRays. The

2>permute 149 312 297f 323 182 300f 44f 274
{many orders
BEST ORDER: 149 312 297f 323 182 300f 44f 274
NEXT BEST : 149 312 297f 323 182 274 44f 300f
LOD vs nextbest = 1.48084299999999

3> evaluate 149 312 297f 323 182 300f 44f 274

NAME	BREAK FREQ	cR
149	0.557	81.4
312	0.485	66.4
297f	0.490	67.3
323	0.494	68.1
182	0.413	53.3
300f	0.347	42.6
44f	0.360	44.6
274	0.000	0.0

LIKELIHOOD = -190.248988
MAP LENGTH = 423.75

4> linked 149 297f LOD=6.88631, THETA=0.471
5> linked 297f 323 LOD=3.743072, THETA=0.540
6> linked 182 300f LOD=5.149929, THETA=0.455
7> linked 300f 44f LOD=7.250421, THETA=0.368
8> linked 44f 274 LOD=7.093646, THETA=0.380

9> set RETENTION_FREQUENCY 0.1
10> evaluate 149 312 297f 323 182 300f 44f 274

NAME	BREAK FREQ	cR
312	0.673	111.8
149	0.708	123.1
297f	0.672	111.5
323	0.757	141.5
182	0.612	94.7
300f	0.492	67.7
44f	0.496	68.5
274	0.000	0.0

LIKELIHOOD = -184.134124
MAP LENGTH = 718.74

Figure 6.21   *RHMAPPER evaluation of the mouse chromosome 2 data.*

accumulative distance for a framework marker and the markers in the subsequent placement interval is approximately the distance between the two framework markers. The placement markers within an interval are ordered only roughly. A distance of zero between two placement markers indicates only that they have the same calculated distance to the previous framework marker. Finally, the placement map is saved in a file called 'fw.map'.

4.2.4   Example 2: Mouse Chromosome 2

To demonstrate the performance of RHMAPPER using a smaller dataset, the mouse dataset used in the RHMAP section was analysed as follows. The mouse directory was created; all markers were found to be in one linkage group, but no framework could be created because the markers were too far apart. Initially, the only parameter changed is the retention frequency, which is set to 0.25. Changing some of the thresholds used by find_triples and assemble_framework1 did not help in this example. The number of markers is 'just' low enough that all permutations can be evaluated without waiting too long (approximately 1 hour) on a Dec Alpha 3000 with 64 megabytes of main memory. Referring to Figure 6.21, the output of the permute command is shown. As with RHMAP, the best order has the first two markers in the wrong order. Commands 4–8 show the two-point lod scores of some marker pairs for comparison with RHMAP scores. As shown in commands 9 and 10, changing the retention frequency from 0.25 to 0.1 influences the distance by a roughly proportional amount.

4.2.5   Customizing RHMAPPER

RHMAPPER can be customized easily by adding Perl commands and new fields to the database. For example, at the Sanger Centre, a command was added to find totally linked markers and to assign a canonical marker to represent the group. The command added a field to the database called 'TL' and can be be viewed, as shown in Figure 6.22.

This example is from a Sanger Centre chromosome-specific database. Marker 10874 is the canonical marker for totally linked group 4; marker 10926 is a marker in this group, etc.

A disadvantage of RHMAPPER is that a user can gain maximal benefit from it only if they know Perl. An advantage is that instead of having to learn an obscure language used only for this purpose, by learning Perl the user has the knowledge of a very powerful language plus maximal benefit of RHMAPPER.

```
3> gd -TL -RHVECTOR 10874 10926 10710
NAME  TL  RHVECTOR
10874  TL4 1001101001000111011111010001101110111001001000011011110001111000100000110001
10926    4   1001101001000111011111010001101110111001001000011011110001111000100000110002
10710    4   1021101001000111011111010001101110111001201000011011112001111000100000110001
etc
```

Figure 6.22 *RHMAPPER has been customized for the Sanger Centre. In this example, the field 'TL' has been added to the database to contain totally linked calculations.*

4.2.6   Using RHMAPPER Via the Web

Markers typed on the human GB4 WG-RH panel can be remotely mapped against the Whitehead RH map at the following web site: http://www-genome.wi.mit.edu, select 'Map STSs relative to the human radiation hybrid map'. Instructions are available on how to enter your vector results from screening a STS against the GB4 panel. Your STS will be placed on the framework map and the results will be sent back by email.

4.3   RADMAP

MultiMap is a computer program for the automated construction of linkage or RH maps. For RH mapping it uses the computer program RADMAP (Matise *et al.*, 1994) for likelihood computation and distance estimation. The mapping algorithms are essentially the same as those used by MultiMap for linkage mapping, although several new functions have been added since the original version of MultiMap was released. RADMAP's calculations are based on the paper by Chakravarti and Reefer (1992). These are similar to those published by Cox *et al.* (1990) and by Boehnke *et al.* (1991) using an equal retention model. A salient difference is when there are missing data, that is, each hybrid's data are expanded at each missing data point to allow computation of a weighted probability of both a plus (+) or a minus (−) at that data point.

RADMAP is still being improved and updated, so a detailed description of the package in this chapter would be premature. MultiMap is freely available at the Web/FTP site at http://linkage.rockefeller.edu/multimap. It runs on most UNIX systems, and requires the installation of LISP, which can also be obtained via FTP. Detailed instructions are available on the Web site. MultiMap includes

Table 6.1   Summary of the RADMAP data analysis

Marker no.	Markers	Map length	Breakage frequency	cRay
1	M149	0.0	0.675	112.3
0	M312	112.3	0.618	96.1
2	M297F	208.4	0.660	107.7
3	M323	316.1	0.744	136.2
4	M182	452.3	0.585	88.0
5	M300F	540.3	0.467	62.9
6	M44F	603.3	0.473	64.0
7	M274	667.2		

an ancillary program, ColorMap, to produce a graphical display of the map, which runs only on Macintosh computers.

An analysis of the mouse chromosome 2 dataset using RADMAP generated the map in Table 6.1. The correct marker order according to the genetic map is ranked 5th with a lod 1.05 difference between it and the best order. The 8 markers on the map are numbered 0 to 7.

5   CONCLUSIONS

The rapidly growing utilization of radiation hybrids for long-range genome mapping has meant that the datasets are now outgrowing some of the less recent RH analysis packages. Thousands of ESTs (expressed sequence tags) are currently being localized on the human genome map using radiation hybrid mapping (Schuler *et al.*, 1996). The successful WG-RH mapping of the human genome has led the way for WG-RH mapping of model organism genomes (Schmitt *et al.*, 1996; McCarthy *et al.*, 1997). WG-RH panels are currently available for mouse, rat, baboon and pig, and work is in progress on dog, chick and zebrafish panels (all unpublished to date), and large-scale EST mapping projects are about to begin using the mouse and rat panels. WG-RH mapping is ideal for mapping model organisms with few existing genetic mapping resources, particularly for larger organisms such as cow and horse, whose long gestation period and single births make meiotic map generation both difficult and time consuming. The advantages of an unlimited source of DNA, the ability to map non-polymorphic markers and the efficiency of long-range map generation by screening fewer than 100 hybrids assure the contribution of WG-RH mapping to physical mapping in model organisms.

When choosing a RH mapping package, the eventual size of the dataset is an important consideration. The RHMAP program has been used widely for many years for mapping small numbers of markers. The expansion in the use of RH mapping led to the development of the RHMAPPER program which accommodates thousands of markers but does not use the various retention models provided by RHMAP. RADMAP is another recently developed program which has been used to build large framework maps.

We analysed eight markers from the mouse genome using each of these three programs. All three analysis programs generated the same best marker order. The output from the RHMAPPER, RHMAP and RADMAP analyses is summarized in Table 6.2. RHMAP and RADMAP calculate similar, and consistently larger, breakage frequencies and distance estimates across the map, compared with RHMAPPER. As well as the breakage frequency and distance in $cRays_{3000}$, the distance between markers is given as a proportion of the overall map length. This proportional distance is almost identical for RHMAP and RADMAP. RHMAPPER generates similar but not identical values. The high similarity

Table 6.2 Comparison of the analysis output from RHMAPPER, RHMAP and RADMAP programs

Marker	RHMAPPER			RHMAP			RADMAP		
	Breakage frequency	cRay	Proportional distance	Breakage frequency	cRay	Proportional distance	Breakage frequency	cRay	Proportional distance
149	0.557	81.4	0.192	0.673	111.6	0.169	0.675	112.3	0.168
312	0.485	66.4	0.157	0.614	95.3	0.144	0.618	96.1	0.144
297f	0.490	67.3	0.159	0.656	106.6	0.161	0.660	107.7	0.161
323	0.494	68.1	0.161	0.740	134.7	0.204	0.744	136.2	0.204
182	0.413	53.3	0.126	0.580	86.8	0.131	0.585	88.0	0.132
300f	0.347	42.6	0.101	0.462	62.1	0.094	0.467	62.9	0.094
44f	0.360	44.6	0.105	0.479	63.3	0.096	0.473	64.0	0.096
274									
Total		423.7			660.4			667.2	

between RHMAP and RADMAP results is to be expected, because they both use similar likelihood calculations under the equal retention model (see Section 4.3). This set of markers spans an approximately 88 Mb region, so for this map, 1 $cRay_{3000} = 208$ kb for RHMAPPER and 1 $cRay_{3000} = 133$ kb for RHMAP.

The variety of analyses and the choice of retention models make RHMAP a very useful analysis package for smaller datasets. However, for the large-scale long-range mapping projects, RHMAPPER and RADMAP would be the programs of choice. An ideal next-generation mapping program would combine the advantages of each of these packages, allowing the application of different analysis methods to large datasets for the purpose of ordering markers and estimating distances, in a program with a graphical interface for ease of map presentation.

ACKNOWLEDGEMENTS

The authors thank Tara Cox Matise for providing information and analysis results for her RADMAP package. We also thank Panos Deloukis and Richard Durbin for comments on the manuscript.

REFERENCES

Baum LE (1972) An inequality and associated maximization technique in statistical estimation for probabilistic functions of Markov processes. *Inequalities* **3**, 1–8.

Boehnke M, Lange K, and Cox DR (1991) Statistical methods for multipoint radiation hybrid mapping. *Am. J. Hum. Genet.* **49**, 1174–1188.

Bouzyk M, Bryant SP, Schmitt K, Goodfellow PN, Ekong R and Spurr NK (1996) Construction of a radiation hybrid map of chromosome 9p. *Genomics* **34**, 187–192.

Burmeister M, Kim S, Price E, de Lange T, Tantravahi U, Myers R and Cox D (1991) A map of the distal region of the long arm of human chromosome 21 constructed by radiation hybrid mapping and pulsed-field gel electrophoresis. *Genomics* **9**, 19–30.

Cox D, Burmeister M, Price ER, Kim S and Myers RM (1990) Radiation hybrid mapping: a somatic cell genetic method for constructing high-resolution maps of mammalian chromosomes. *Science* **250**, 245–250.

Chakravarti A and Reefer JE (1992) A theory for radiation hybrid (Goss–Harris) mapping: application to proximal 21q markers. *Cytogenet. Cell Genet.* **59**, 99–101.

Dausset J, Ougen P, Abderrahim H, Billault A, Sambucy JL, Cohen D and Le Paslier D (1992) The CEPH YAC library. *Behring Inst. Mitt.* 13–20.

Dempster AP, Laird NM and Rubin DB (1977) Maximum likelihood from incomplete data via the EM algorithm. *J. R. Stat. Soc. B.* **39**, 1–22.

Frazer K, Boehnke M, Budarf ML, Wolff RK, Emanuel BS, Myers RM and Cox DR (1992) A radiation hybrid map of the region on human chromosome 22 containing the neurofibromatosis type 2 locus. *Genomics* **14**, 574–584.

Goss S and Harris H (1975) New method for mapping genes in human chromosomes. *Nature* **255**, 680–684.

Gyapay G, Schmitt K, Fizames C, Jones H, Vega-Czarny N, Spillett D, Muselet D, Prud'Homme J-F, Dib C, Auffray C, Morissette J, Weissenbach J and Goodfellow PN (1996) A radiation hybrid map of the human genome. *Hum. Mol. Genet.* **5**, 339–346.

Hudson TJ, Stein LD, Gerety SS, Ma J, Castle AB, Silva J, Slonim DK, Baptista R, Kruglyak L, Xu SH, Hu XT, Colbert AME, Rosenburg C, Reevedaly MP, Rozen S, Hui L, Wu XY, Vestergaard C, Wilson KM, Bae JS, Maitra S, Ganiatsas S, Evans CA, Deangelis MM, Ingalls KA, Nahf RW, Horton LT, Anderson MO, Collymore AJ, Ye WJ, Kouyoumjian V, Zemsteva IS, Tam J, Devine R, Courtney DF, Renaud MT, Nguyen H, O'Connor TJ, Fizames C, Faure S, Gyapay G, Dib C, Morissette J, Oorlin JB, Birren BW, Goodman N, Weissenbach J, Hawkins TL, Foote S, Page DC and Lander ES (1995) An STS-based map of the human genome. *Science* **270**, 1945–1954.

James MR, Richard CW III, Schott JJ, Yousry C, Clark K, Bell J, Terwilliger JD, Hazan J, Dubay C, Vignal A, Agrapart M, Imai T, Nakamura Y, Polymeropoulos M, Weissenbach J, Cox DR and Lathrop GM (1994) A radiation hybrid map of 506 STS markers spanning human chromosome 11. *Nature Genet.* **8**, 70–76.

Keats BJB, Shermsan SL, Morton NE, Robson EB, Buetow KH, Cartwright PE, Chakravarti A, Francke U, Green PP and Ott J (1991) Guidelines for human linkage maps – an international system for human linkage maps. *Genomics* **9**, 557–560.

Kumlien J, Grigoriev A, Ross M, Roest Crollius H, Keil H, Goodfellow PN and Lehrach H (1994) Construction of a radiation fusion hybrid map over the X chromosome. *Cytogenet. Cell Genet.* **67**, 341.

Lander ES, and Green P (1991) Counting algorithms for linkage – correction to Morton and Collins. *Ann. Hum. Genet.* **55**, 33–38.

Lange K, Boehnke M, Cox DR and Lunetta KL (1995) Statistical methods for polyploid radiation hybrid mapping. *Genome Res.* **5**, 136–150.

Leach RJ and O'Connell P (1995) Mapping of mammalian genomes with radiation (Goss and Harris) hybrids. *Adv. Genet.* **33**, 63–99.

Lunetta KL, Boehnke M, Lange K and Cox DR (1996) Selected locus and multiple panel models for radiation hybrid mapping. *Am. J. Hum. Genet.* **59**, 717–725.

Matise TC, Perlin M and Chakravarti A (1994) Automated construction of genetic linkage maps using an expert system (MultiMap): a human genome linkage map. *Nature Genet.* **6**, 384–390.

McCarthy L (1996) Whole genome radiation hybrid mapping. *Trends Genet.* **12**, 491–493.

McCarthy LC, Terrett J, Davis, ME *et al.* (1997) A first generation whole genome-radiation hybrid map spanning the mouse genome. *Genome Res.* **12** (in press).

Raeymaekers P, Van Zand K, Jun L, Hoglund M, Cassiman JJ, Van den Berghe H and Marynen P (1995) A radiation hybrid map with 60 loci covering the entire short arm of chromosome 12. *Genomics* **29**, 170–178.

Richard CD, Withers D, Meeker T, Maurer S, Evans G, Myers R, and Cox D (1991) A radiation hybrid map of the proximal long arm of human chromosome 11 containing the multiple endocrine neoplasia type 1 (MEN-1) and bcl-1 disease loci. *Am. J. Hum. Genet.* **49**, 1189–1196.

Richard CD, Boehnke M, Berg D, Lichy J, Meeker T, Hauser E, Myers R and Cox D (1993) A radiation hybrid map of the distal short arm of human chromosome 11, containing the Beckwith–Wiedemann and associated embryonal tumor disease loci. *Am. J. Hum. Genet.* **52**, 915–921.

Schmitt K, Foster JW, Feakes RW, Knights C, Davis ME, Spillet DJ and Goodfellow PN (1996) Construction of a mouse whole genome radiation hybrid panel and application to MMU11. *Genomics* **34**, 193–197.

Schuler G, Boguski MS, Stewart EA, Stein LD, Gyapay G, Rice K, White RE, Rodrigueztome P, Aggarwal A, Bajorek E, Bentolila S, Birren BB, Butler A, Castle AB, Chiannilkulchai N, Chu A, Clee C, Cowles S, Day PJR, Dibling T, Drouot N, Dunham I, Duprat S, East C, Edwards C, Fan JB, Fang N, Fizames C, Garrett C, Green L, Hadley D, Harris M, Harrison P, Brady S, Hicks A, Holloway, E, Hui L, Hussain S, Louisditsully C, Ma J, Macgilvery A, Mader C, Maratukulam A, Matise TC, McKusick KB, Morissette J, Mungall A, Muselet D, Nusbaum HC, Page DC, Peck A, Perkins S, Piercy M, Qin F, Quackenbush J, Ranby S, Reif T, Rozen S, Sanders C, She X, Silva J, Slonim DK, Soderlund C, Sun WL, Tabar P, Thangarajah T, Vegaczamy N, Vollrath D, Voyticky S, Wilmer T, Wu X, Adams MD, Auffray C, Walter NAR, Brandon R, Dehejia A, Goodfellow PN, Houlgatte R, Hudson JR, Ide SE, Iorio KR, Lee WY, Seki N, Nagase T, Ishikawa K, Nomura N, Phillips C, Polymeropoulos MH, Sandusky M, Schmitt K, Berry R, Swanson K, Torres R, Venter JC, Sikela JM, Beckman JS, Weissenbach J, Myers RM, Cox DR, James MR, Bentley D, Deloukas P, Lander ES and Hudson TJ (1996) A gene map of the human genome. *Science* **274**, 540–546.

Shaw SH, Farr JE, Thiel BA, Matise TC, Weissenbach J, Chakaravarti A and Richard CWR (1995) A radiation hybrid map of 95 STSs spanning human chromosome 13q. *Genomics* **27**, 502–510.

Slonim D, Kruglyak L, Stein L, and Lander E (1996) Building human genome maps with radiation hybrids. RECOMB 97, ACM Press.

Stein L, Kruglyak L, Slomin D, and Lander E (1995) RHMAPPER software package http://www-genome.wi.mit.edu/ftp/pub/software/rhmapper.

Stewart E, McKusick K, Aggarwal A, Bajorek E, Brady S, Chu A, Fang N, Hadley D, Harris M, Hussain S, Lee R, Maratukulum A, O'Connor K, Perkins S, Piercy M, Qin F, Reif T, Sanders C, She X, Sun W, Tabar P, Voyticky S, Cowles S, Fan J, Mader C, Quackenbush J, Myers R and Cox D (1997) An STS-based radiation hybrid map of the human genome. *Genome Res.* **7**, 422–433.

Walter M and Goodfellow P (1993) Radiation hybrids: irradiation and fusion gene transfer. *Trends Genet.* **9**, 352–356.

Walter M, Spillet D, Thomas P, Weissenbach J and Goodfellow P (1994) A method for constructing radiation hybrid maps of whole genomes. *Nature Genet.* **7**, 22–28.

Winokur ST, Schutte B, Weiffenbach B, Washington SS, McElligott D, Chakravarti A, Wasmuth JH and Altherr MR (1993) A radiation hybrid map of 15 loci on the distal long arm of chromosome 4, the region containing the gene responsible for facioscapulohumeral muscular dystrophy (FSHD). *Am. J. Hum. Genet.* **53**, 874–880.

7 Sequence Ready Clone Maps

Carol Soderlund, Simon Gregory and Ian Dunham

The Sanger Centre, Wellcome Trust Genome Campus, Hinxton, Cambridge CB10 1SA, UK

1 INTRODUCTION

Determination of the human genome DNA sequence will provide the informational infrastructure for human biology and medicine in the twenty-first century. With incremental improvements in technology, the focus has shifted from whether this goal can be achieved to the practicalities of time, cost and accuracy. Thus we are now in the production stage of human genomics and our strategic concerns are with efficiency. It is in this context that we describe our strategy for producing substrate for human genome sequencing, and the associated software.

High-throughput DNA sequencing using shotgun dideoxy-chain termination biochemistry and automated gel electrophoresis with laser fluorescence detection requires a constant supply of M13 or plasmid template DNAs. For small genomes or chromosomes, it is possible to produce this template by shotgun cloning of the whole genome or chromosome (Fleischmann *et al.*, 1995). However, for the human genome, the length of DNA involved and the presence of dispersed repetitive sequences, makes this approach impractical. Instead, most mammalian genomic sequencing uses an intermediate cloning step in which the genome or chromosome is cloned in pieces of between 30 and 250 kb in a bacterial cloning vector such as a cosmid, P1 artificial chromosome (PAC) (Ioannou *et al.*, 1994) or bacterial artificial chromosome (BAC) (Shizuya *et al.*, 1992) to construct a clone map or 'contig' of the region (Figure 7.1). Representative clones from the map (the minimal tiling path or spanning set) are then used as the DNA source for the shotgun M13 or plasmid library, and the sequence of the genome is assembled from the shotgun sequence reads in a series of cosmid, PAC or BAC projects.

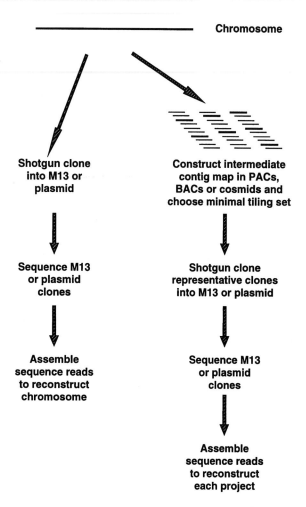

Figure 7.1 *Shotgun sequencing of a whole chromosome versus sequencing via an intermediate map. On the left hand side is illustrated the whole chromosome shotgun strategy. On the right hand side, genomic sequencing via a sequence ready map strategy is illustrated. The small horizontal lines represent bacterial clones containing genomic DNA, overlapping to form a contig. The thicker lines are clones that have been chosen as the representative set forming the minimal tiling path.*

The bacterial clone map strategy has the advantage of keeping each sequencing project relatively small, self-contained and compatible with current sequence assembly algorithms and software, and allows effective integration of the projects with a series of well established genomic technologies (see below). On the other hand, working with a series of bacterial clone projects introduces a number of what can be termed 'mapping' issues. It is important to know the neighbours of each sequencing clone, and the extent of overlap between the adjacent projects. Excessive overlap between the projects leads to increased resequencing and increased cost. Therefore, the aim is to choose the clones to be sequenced with a minimal overlap to ensure continuity of the genomic sequence. To give the best choice for the minimal tiling path, it is optimal to have a clone map that is contiguous over several megabases, and has multiple independent clones covering all points (dense/deep/multifold coverage). Ideally the clone map is constructed well in advance of the sequencing project as in the cases of *Saccharomyces cerevisiae* (Riles *et al.*, 1993) and *Caenorhabditis elegans* (Coulson *et al.*, 1986). However, with the rush to sequence large regions of the human and other genomes, clone maps have not necessarily been established, and it is necessary to construct the map and sequence almost in parallel. In this situation, the contig must also be verified as sequencing continues by integration with other maps of the genome or chromosome.

In summary, these are the properties required of the 'sequence ready' clone map.

- Several-fold coverage allows identification of the clones forming an optimal minimal tiling set with overlaps of small but defined size. This depth of contig also gives security against deletions in individual clones and erroneous contig construction.
- The contigs should be constructed with a method that allows integration of clones from different libraries, as improvements in technology will continue.
- Each clone must have a unique name and a set of experimental data attached to that name, at least some of which can be linked directly to its sequence. This enables subsequent tracking of a clone through the sequencing process.
- The map should be anchored to genome or chromosome-wide, long-range maps in order to coordinate within the project and with the rest of the world. In this chapter we refer to this type of map alternatively as the framework map or the long-range marker map.

2 STRATEGY

A number of strategies have been proposed for rapid production of sequence ready clone maps (Roach *et al.*, 1995; Kim *et al.*, 1996; Venter *et al.*, 1996). None of these strategies has so far been demonstrated to work successfully in a large-scale production environment. The strategy that we have chosen to adopt has been developed by incremental change from existing methods, and is illustrated in Figure 7.2.

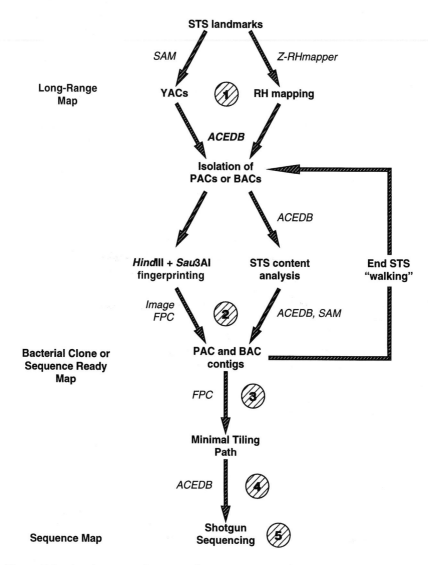

Figure 7.2   *Our integrated strategy for construction of sequence ready clone maps. The software elements of the strategy are indicated in italics. The circled numbers refer to the stages involved in process tracking and coordination (see Section 7.2).*

For our human genome sequencing projects, we work on a chromosome by chromosome basis because this is the manageable unit for both coordination and experimental approach. To coordinate the sequence ready contig construction we use a long-range map of the chromosome comprising sequence tagged site (STS) markers which have been ordered by a method such as radiation hybrid

(RH) mapping (Cox *et al.*, 1990; Gyapay *et al.*, 1996) or yeast artificial chromosome (YAC) mapping (Green and Olson, 1990). We aim to use a map that has of the order of 15 to 20 markers per megabase as this will allow local ordering of bacterial clone contigs. Thus the main criteria for the long-range map are that it must be reliable in long-range ordering and have a resolution of less than a megabase. The markers in the long-range map are then used to isolate bacterial clones from whole genome BAC or PAC libraries. The average size of the human genomic DNA contained in these clones is between 100 and 150 kb. The clones isolated are analysed by a high-throughput fingerprinting method based on a *Hin*dIII/*Sau*3AI digest in which all the *Hin*dIII sites are labelled (Coulson *et al.*, 1986). The information generated for each clone is a set of fragments or 'bands' characteristic of the DNA content of the clone and such that the number of bands is twice the number of *Hin*dIII sites in the clone, which in turn is on average related to the size of the clone. Contigs of clones are assembled by comparing all the sets of fragments and determining groups of clones that are likely to overlap based on sharing a significant number of bands. Additional information is used to aid the assembly by determining the STS content (Green and Olson, 1990) of all the clones. Not only does the STS content provide confirmation of overlap, but it also places the nascent contigs relative to the framework map. After completion of the initial clone isolation and contig assembly, we have a set of contigs placed and oriented relative to the long-range map. Closure of the gaps between contigs is achieved by development of new markers from the ends of each contig either as STSs after sequencing of the ends of PAC or BACs, or as DNA hybridization probes by one of a number of 'end-rescue' strategies (Riley *et al.*, 1990; Wu *et al.*, 1996). As soon as the contigs achieve sufficient coverage and extent to be reliable, it is possible to pick tiling paths for genomic sequencing by identification of the clones with fingerprint and STS content data consistent with the contig assembly, and sufficient minimal overlap with its neighbour. However, if the aim is to obtain a truly minimal overlapping set for sequencing, it is best to wait for substantial contigs to be formed because tiling paths will have to be adjusted each time two contigs are joined.

This method has already been implemented successfully in construction of bacterial clone contigs of over 80 Mb in our projects for sequencing human chromosomes 1, 6, 20, 22 and X at the Sanger Centre. Hence, in the remainder of this chapter we will describe the software that has been developed to enable effective construction of sequence ready maps under this scheme.

3 SOFTWARE FOR SEQUENCE READY MAPS

Table 7.1 summarizes the software components that are used at each stage of our strategy. For each stage we will describe how the programs work, how to obtain the software and how to use the programs with some real examples.

Table 7.1 Software used for sequence ready contigs

Map task	Method	Software component
1. Long-range map	Radiation hybrids, YACs	Z-RHMAPPER, SAM
2. BACs/PACs STS content	PCR and hybridization	ACEDB, SAM
3. Overlap map of BACs/PACs	*Hind*III/*Sau*3AI fingerprinting	Image, FPC
4. Integration of 1, 2 and 3	Software based	FPC
5. Choice of minimal tiling set	Manual	FPC
6. Tracking	Software based	ACEDB
7. Coordination	System cronjobs	Use of common .ace format

Finally we will describe how we link the components together to enable effective coordination of the different stages of the strategy. Some of the programs involved have been extensively described elsewhere, and in these cases we will only place them into the context of our overall strategy, and reference to where further information can be found. In our system the software has been implemented on a network of Unix machines (SunOS, Solaris, DEC Alpha, SGI). While it may be possible to construct similar systems under other operating systems, we have not tested this, and would advise the use of Unix platforms.

4   LONG-RANGE MAP CONSTRUCTION (SAM, Z-RHMAPPER)

Long-range maps can be created by YAC/STS mapping, RH mapping, or from published data.

4.1   Building YAC/STS Maps with SAM

Given a set of clones and their marker content information, the markers can be ordered by finding the permutation that maximizes the number of consecutive markers per clone over all the clones. Due to false-negative and false-positive scores, the problem is NP complete; that is, to guarantee finding the best order of the markers, all permutations have to be scored and this takes an unacceptable amount of time for large datasets, so an approximation algorithm must be used. We use the program SAM (System for Assembling Markers) to assemble, update, query and edit the marker maps (Soderlund and Dunham, 1995). Our largest marker assembly project is the chromosome 22 markers ordered on YACs (Collins *et al.*, 1995). We currently have 844 markers ordered on 778 YACs for a resolution of one marker per 50 kb. This map is used as our long-range map for assembly of bacterial clone contigs. Details of how to use SAM are provided in the manual (Soderlund, 1995). The following will briefly describe the scenario we used for building this map.

The input to SAM is a marker file containing the clone names and associated marker content. Multiple biologists work on the project by each taking a region of the chromosome and building a map in a region around one or more genetic markers. The contigs are saved in solution files which can be read, written, edited and merged. A solution file for a region is started by using the `Follow Marker` function which brings in all clones that score positive for a given marker, such as a specified genetic marker. The clones and associated markers are assembled and then the map is extended by using `Follow Marker` for markers found on the end of the contig. The assembly algorithm can run on regions of the map, so only the extended region need be assembled.

New marker content data are daily entered into ACEDB (Durbin and Thierry-Mieg, 1994). An automated script is run nightly which dumps the data from ACEDB and updates the SAM marker file using a program called aXXs. When a user reads in a solution file, it is automatically updated, as shown in Figure 7.3. New markers are placed at the end of the map and can be positioned in the existing map using the `slide` function. Solutions can be written in an ACEDB file format and are loaded into the chromosome-specific ACEDB.

It is also practical to use SAM to view the results of STS content analysis on bacterial clones. The STS content data that are generated in ACEDB and exported to FPC (see Sections 5 and 6.3 below) can also be assembled independently using SAM, and the results compared with the results of the fingerprinting approach. In some cases STSs at the ends of clones can reveal overlaps that are not detected at high probability in the fingerprint data, and analysis in SAM can reveal this.

The chromosome 22 solution files are available from ftp.sanger.ac.uk/ /pub/human/chr22/release-1996-11-25/sam. The SAM source, manual and executables are available from ftp.sanger.ac.uk//pub/sam.

4.2   Building Radiation Hybrid Maps using Z-RHMAPPER

RH mapping involves scoring markers against an RH panel (Cox *et al.*, 1990) and comparing the resulting patterns of positive and negative hybrids (indicated by '1' or '0' respectively). Markers are ordered based on the similarity of their patterns, commonly referred to as 'vectors'. A RH can have multiple fragments from different parts of the genome, and therefore from different parts of the chromosome being mapped. Although a good ordering generally has a maximal

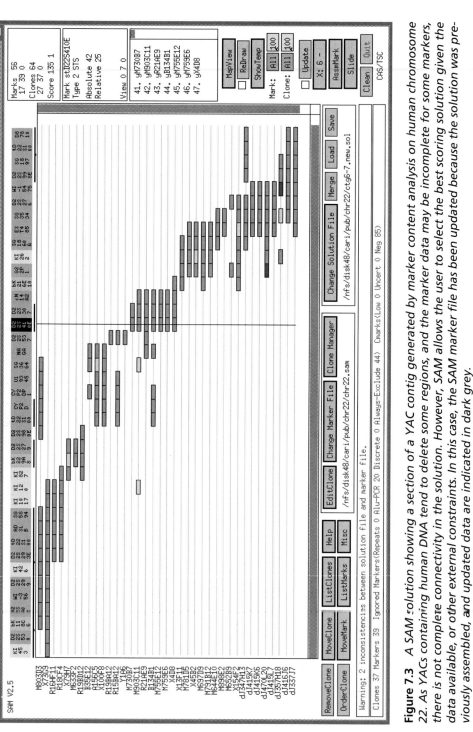

Figure 7.3 A SAM solution showing a section of a YAC contig generated by marker content analysis on human chromosome 22. As YACs containing human DNA tend to delete some regions, and the marker data may be incomplete for some markers, there is not complete connectivity in the solution. However, SAM allows the user to select the best scoring solution given the data available, or other external constraints. In this case, the SAM marker file has been updated because the solution was previously assembled, and updated data are indicated in dark grey.

number of consecutive hybrid markers (as in the YAC/STS maps), there are generally multiple chromosomal marker groups per hybrid which complicates the computation. A benefit of this method is that it gives an estimate of the distance between markers. We use RHMAPPER (Stein *et al.*, 1995; Slonim *et al.*, 1996) to order the markers and estimate the distance. RHMAPPER was designed to accommodate large numbers of markers and has been used at the Whitehead Institute/MIT Center for Genome Research to build a whole genome map (Hudson *et al.*, 1995). We have extended it to accommodate dense chromosome maps and our dataflow. The modified version is referred to as Z-RHMAPPER. The following briefly describes our scenario (see Chapter 6 for more information on RH mapping).

Markers are scored against the Genebridge4 panel (Gyapay *et al.*, 1996) and the results are entered into an ACEDB database. The data are dumped from ACEDB into the data format to be loaded into the RHMAPPER database. The initial frameworks were created by two different methods: (1) the genetic order was verified and edited using various RHMAPPER functions; and (2) the framework was taken from the published genome map (Schuler *et al.*, 1996). The function Ztl is run to find totally linked groups, where totally linked markers have identical vectors (ambiguous data match '0' or '1'). A marker is picked from each group, either a framework marker or one that has no ambiguous data, to be the canonical marker to represent the totally linked group. The framework is extended by using the grow_framework function with the canonical markers as input. All remaining markers are binned within the framework using the placement_map function. The resulting map is written to a file which is subsequently read into ACEDB and displayed on our WWW site (http://www.sanger.ac.uk/HGP/Rhmap/).

5   STS-CONTENT DATA STORAGE USING ACEDB

Each clone that is included in the bacterial clone map has been identified as positive for a particular STS landmark. This may be by polymerase chain reaction (PCR) testing the clone, or by DNA–DNA hybridization. This information anchors the clone to a particular position in the long-range framework map and, as contigs grow and have landmark information from a larger region of the framework map, provides an orientation for the contig on the chromosome. In addition the STS content of a set of clones gives confirmation of the clone assembly from the fingerprints, because it is to be expected that clones containing the same landmarks should also overlap by their fingerprints. Thus, we accumulate the landmark information for all clones.

The STS content information for the bacterial clones is entered and stored in an ACEDB database through a user interface known as the Grid. The single clones in each library are organized and stored in unique well addresses in microtitre plate arrays. Positive PCR or hybridization data can be ascribed to a clone by clicking on that position of a representation of the microtitre plate grid. The ACEDB program has a number of tools available for data entry and display of these data. Our use of ACEDB for clone library data management has been extensively described previously (Dunham *et al.*, 1994; Dunham and Maslen, 1996). Details of how to obtain the software, and its implementation are available in these references and also at http://www.sanger. ac.uk/Software.

6   SEQUENCE READY CONTIG CONSTRUCTION USING FPC

6.1   Methodology

A fingerprint of a clone can be a set of fragments, where the value of a fragment is its relative migration distance or its length. If two fragments have the same value, they may be the same. As the measurement is not exact, the user must set a tolerance. For migration values, if two fragments are within +/− of the tolerance, the fragments may be the same. For fragment lengths, the tolerance is computed based on the fragment length. FPC will work for either type of fingerprint.

If two clones have one or more fragments within the tolerance, they may overlap. The number of shared fragments between two clones is counted, and the probability that the shared fragments arise by coincidence is computed. The user supplies a cutoff and, if the coincidence score is below the cutoff (i.e. the probability that the shared bands arise by chance is low), the clones are said to overlap. If the cutoff is set too stringently, many truly overlapping clones will be ignored (false negatives), and if it is set at a low stringency, clones will be accepted as overlapping when they are not (false positives).

If clones A and B overlap, the fragments can be ordered into three partially ordered groups g1, g2, and g3 (Figure 7.4a). The order of the groups cannot be changed but the order of the fragments within a group is not known. If clone A overlaps clones B and C, but the overlap with clone C is not detected, this false negative can be resolved by ordering the fragments (Figure 7.4b). This shows that clones A and C overlap by two fragments. We call this type of layout of the fragments a consensus bands (CB) map, because the actual value of each band is the consensus of all the bands in the column. Finding the optimal order of the fragments is the same problem as finding the optimal order of STSs, except that for STS–YAC mapping the set of STSs is known whereas for building a CB map the set of fragments is not known and must be approximated, which complicates the computation.

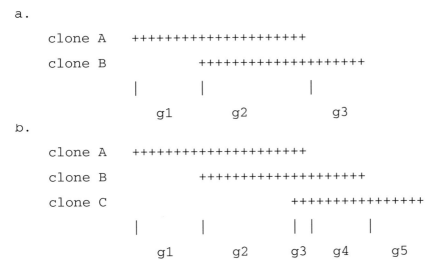

Figure 7.4 *The rationale of fingerprint mapping and the consensus band (CB) map. A '+' is a fragment and multiple '+' in a column indicate shared fragments.*

6.2 Generating the Fragment Sets

We have utilized two separate methods to generate characteristic fragments sets of bacterial clones. One method uses a radioactive reporter molecule incorporated in the 'sticky end' of *Hin*dIII digested cloned insert DNA (Coulson *et al.*, 1986; Gregory *et al.*, 1997), and the second is a modification that makes use of fluorescent dye terminators for end labelling (S.G. Gregory *et al.*, unpublished results). Although both methods use a polyacrylamide gel matrix to separate fragment sets, our fluorescent method uses the Perkin–Elmer Applied Biosystems 377 for the generation of fingerprint data. The raw fluorescent gel image is transferred from the ABI 377 and ported into the UNIX-based Image program, which is used in a semi-automated stepwise manner. Image is used to transfer the fingerprint data from fluorescent band patterns on the raw gel image to a set of numerical values. Within Image clone names and gel number are attached to the raw gel image, and the lanes containing the restriction digests are tracked along the length of the gel. The four separate wavelengths present in the gel image are deconvoluted, and the fingerprint bands can be edited, enabling background or partial bands to be removed from the band set. Finally the marker standard, which is used to assign standardized migration values, is locked with the established standard band set. After these stages have been completed, the fingerprint information for each clone can be added to the relevant fingerprinted contigs FPC project (see below). An introduction to Image, installation and documentation can be found at http://www.sanger.ac.uk/Software/Image.

6.3   Using FPC V2.8

6.3.1   Overview

The contig9 software (Sulston *et al.*, 1988) was developed for building fingerprint contigs for *C. elegans*. FPC (Figure 7.5) is a second-generation program which utilizes most of the same concepts as contig9 and has additional functionality

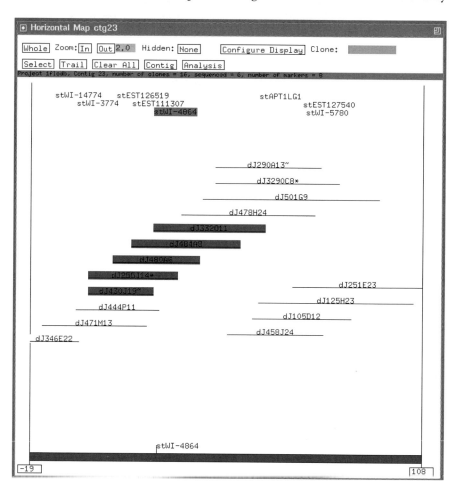

Figure 7.5   *FPC. The contig display is divided into four horizontal tiers for markers, clones, comments and anchor point, respectively. The characters {∗, =, ~, #} represent canonical, equal, approximate and pseudo-buried, respectively. Buried clones are generally positioned with their left ends at the left end of their canonical clone. The marker stWI-4864 has been clicked on and the clones for which it is positive are highlighted.*

to support large-scale fingerprinting, marker data, long-range markers and the selection of sequence ready clones. FPC uses an ACEDB-style flat format as its database and employs the ACEDB graphics library (Durbin and Thierry-Mieg, 1994).

FPC uses the probability of coincidence score developed for contig9 and the concept of the *canonical* clone. If a clone has a subset of another clone's bands, it can be *buried* in the larger clone. A canonical clone can have many buried clones but cannot itself be buried. A clone can be buried as 'exact' or 'approximate' if all or most of the bands are the same, respectively. FPC has a third type of buried clone called 'pseudo', which does not require an approximate match but allows the user to bury a clone in its approximate position in order to limit the depth and/or ignore poorly fingerprinted clones. Buried clones are not displayed unless requested.

A marker is automatically displayed with a contig if it belongs to one or more clones in the map and is positioned at the midpoint of the largest stack of clones containing the marker. A marker can be defined as an anchor point, and all anchor points are always displayed along the bottom of the contig to provide framework information when a region of the contig is displayed. Our long-range markers are automically defined as anchor points.

Details of FPC are provided in the manual (Soderlund and Longden, 1996). The details of the algorithm are provided by Soderlund *et al.* (1997). The following is an overview of using FPC. See http://www.sanger.ac.uk/Users/cari/fpc_faq.html for information on new releases and frequently-asked questions.

6.3.2   Installing FPC

FPC can be obtained by anonymous ftp from ftp.sanger.ac.uk, in directory pub/fpc. The software is written in C and should be portable to any Unix machine with a standard C compiler and X windows. The ftp directory also contains a file called Demo.tar. To try FPC, ftp the demo file and an executable (e.g. fpc.ALPHA), and:

```
>tar xvf Demo.tar
tar: blocksize=60
x Demo/Image/16.bands, 4412 bytes, 9 tape blocks
x Demo/demo.cor, 4430 bytes, 9 tape blocks
x Demo/demo.fpc, 13562 bytes, 27 tape blocks
> mv fpc.ALPHA Demo/fpc
> cd Demo
> fpc demo
```

The last line executes FPC with the file called demo.fpc. Two windows will be initiated, the one shown in Figure 7.6 and a message window. These two windows stay active during the entire FPC session and all error, warning and information messages are written in the message window.

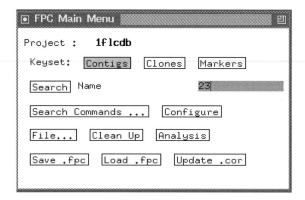

Figure 7.6   *The FPC main window. The project name can be double clicked to show information about the project. Keysets can be displayed by double clicking the corresponding button for contigs, clones and markers. Subsets of a keyset can be created with the* Search Commands... *based on creation date, gel, etc. The pull-down menu* File... *gives access to the site-specific functions along with other file functions such as a* Write Lock *for exclusive write access as there are often multiple people working on a project. After any salient changes, select* Save .fpc *to save the changes.*

6.3.3   Starting an FPC project

As described in Section 6.2, the Image program outputs a file per gel which contains the clone names and bands. These files have a '.bands' suffix and are referred to as the *.bands files where the '*' is usually the gel number.

FPC uses the concept of a project where the files for a project are in the project directory. To start a project, create a directory with a subdirectory called Image. Put the *.bands files in the Image directory. Start up FPC:

- On the File... pull-down menu, select Create new project and enter the project name. For example, to create the demo files, we entered 'demo'.
- Execute Update .cor. The bands files will be read, the clones and bands will be added to the FPC files with the suffixes '.fpc' and '.cor' respectively. The bands files are moved into a directory called Bands. At this point, an 'ls' on the directory demo will show the following:

```
Bands/              Image/              demo.cor
demo.cor.backup     demo.fpc            demo.fpc.backup
demo.fpp
```

The following describes one scenario for building new contigs. FPC offers a wide range of functions, and alternative scenarios can be used for building contigs.

6.3.4 Creating New Contigs

Select `Analysis` from the FPC Main Menu window and the window shown in Figure 7.7 will be initiated.

- Select `SINGLE`. A function will create 0 or more maps where each map contains a set of transitively overlapping clones.
- Select either `CB coords` or `Len=bands`. A contig will be built for each map. A project window will appear listing the new contigs, as shown in Figure 7.8. The clones are positioned automatically and clones that can buried with 'exact' or 'approximate' status will be buried. The `CB coords` gives the clones' coordinates based on their alignment to the CB map. When the data are clean and the contig is not too big, these coordinates are optimal. Otherwise, many of the clone coordinates are poor, although the positioning is generally correct, in which case it is better to use the length of the clone. With the `Len=bands` option, it uses the same position as calculated but extends or reduces the length such that it is equal to the number of bands present in the fingerprint.
- If the cutoff and tolerance are not set optimally, you will get too few or too many contigs, or your contigs will have many clones that do not place well. In this case, go to the `FPC Analysis` window and enter the size of the biggest contig beside `Kill` and execute `Kill`; this

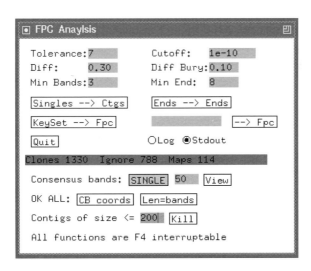

Figure 7.7 *The* `FPC Analysis` *window provides functions for comparing clones against the database, for comparing ends of contigs and for building CB maps for all the singletons (see text).*

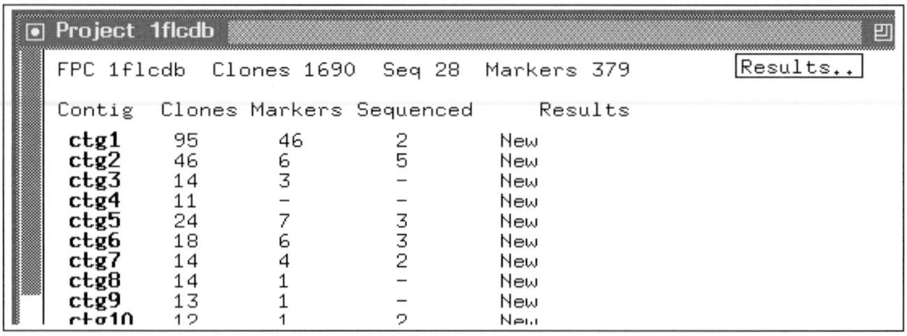

Figure 7.8   *When new contigs are automatically created with the* SINGLE/OK ALL *functions, the new contigs are listed in the project window as shown here. The columns show the contig number, the number of clones in each contig, the number of markers in each contig and the number of clones picked in the sequence tiling path, respectively.*

removes all contigs of the given size and smaller. Alter the parameters and run SINGLE again.

6.3.5   Refining New Contigs

After creating the contigs, verify and refine each new contig. This is done easily by double clicking each contig name in turn from the project window. When a contig is displayed, select the Analysis button, which brings up the window shown in Figure 7.9.

If the map is large, evaluate it in sections by selecting a region (see Section 6.3.6 Editing Contigs), and run the CB algorithm on the selected set. Otherwise, select Calc to run the algorithm on the whole contig. A CB map will be displayed as shown in Figure 7.10.

Evaluate the map by executing Again a few times and observe whether the clones stay in the similar positions. If the maps vary a lot, this indicates a lot of error in the data or some other problem. Clones that have an 'A' over their name may not be positioned well; if the 'A' remains on subsequent Again, it is likely that the clone has a poor fingerprint. Pseudo-bury poor clones or clones in regions of great depth. Then run the CB algorithm again; when the Hidden state is Buried, the buried clones are ignored by the CB algorithm. This may result in multiple maps if a buried clone joins two subcontigs; if this happens, find the offending clone and unbury it. If a given CB map has better coordinates than those in the current contig, select OK to instantiate the new coordinates. Final editing can be done manually, as described in the following section.

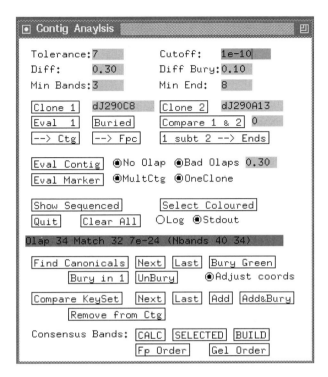

Figure 7.9   *The* Contig Analysis *window is initiated from a contig display. The functions at the top evaluate the current contig. For example, a clone can be compared against the clones in the contig or against the rest of the database. The functions at the bottom suggest clones to add or bury, and performs the edit if requested. The bottom two rows are functions for building CB maps of the contig.*

6.3.6   Editing Contigs

The clone text window in Figure 7.11 shows the values associated with a clone.

By selecting the Edit button, a window is initiated which allows the editing of the clone values. Similarly, a marker can be edited. The position of a clone can also be changed via the Select Option window. A group of clones can be selected by delimiting a region of clones or picking them; various functions operate on the selected set.

A display of the fingerprints in ascending order can be created as shown in Figure 7.11. The fingerprints can be moved, removed and added. If using Image to produce the band files, the user can request that a file of the gel images be put into a format that is read by FPC. Images from different gels can be viewed together in an FPC display (not shown).

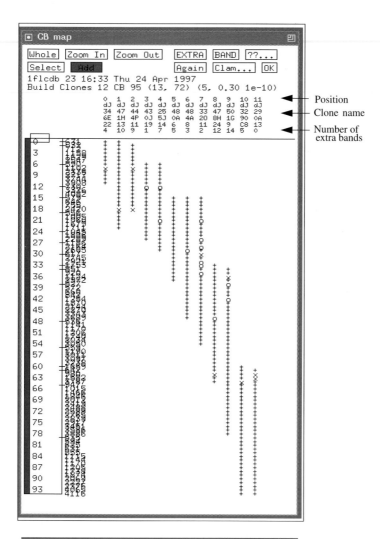

Figure 7.10

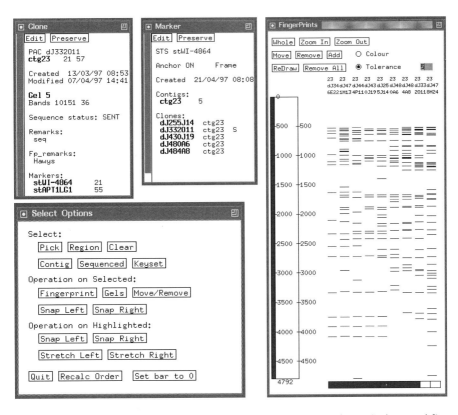

Figure 7.11   *The clone text window, marker text window, select window and fingerprint window.*

Figure 7.10   *A consensus band (CB) map. The numbers along the left are the consensus bands. The row of numbers (i.e. starting with 6 1 * ...) is the position of the clone in the contig (if this is run from the contig window). The '*' indicates that the clone is buried (or could be buried) in the first non-'*' to the left. Below this row is the clone name (which takes up four rows). The last row of this set is the number of bands that could not be placed. Beneath the clone information is a column of {+,o,x} where '+' indicates the clone contains the band, 'o' indicates it does not contain the band, 'x' indicates it contains a band greater than the tolerance but less than twice the tolerance. The* Extra *button displays a window showing the bands that were not placed for each clone. The first time the routine is run, the clones are listed in the order in which they were added to the map. This results in the clones with poorer probabilities being displayed on the right and they may have poor positioning. Each time* Again *is run, an alternative map will be calculated but the order of the clones will not change.*

6.3.7   Adding New Clones to the Database

One or more new *.bands files can be put in the Image directory. Select `Update` `.cor` and the clones and bands will be added to the existing database. To create new contigs from all singletons in the database (new clones and unplaced clones), run `SINGLE/OK ALL`.

    To add clones to existing contigs, create a keyset of the new clones by requesting the clones added after a given date (e.g. yesterday's date). From the `FPC` `Analysis` window, compare the new clones against the rest of the clones in the database by executing `Keyset --> Fpc`. If requested, the output will go to a file. The project window will be displayed listing the number of hits for each contig as shown in Figure 7.12.

    For each contig with hits, select `Compare Keyset` from the `Contig` `Analysis` window. An internal list is created of clones that match the contig. Step through the list by selecting `Next`; a clone is listed and all the clones it matches are highlighted. To add the clone, select `Add`, then either bury the clone or refine its position.

6.3.8   Merging Contigs

On the `FPC Analysis` window, select `Ends --> Ends` and a project window will be displayed as shown in Figure 7.13.

    On the contig window, select `Contig` which allows you to specify the contig to display beside the existing one. The visiting contig can be shifted left or right and the clones can be selected such that they can be displayed in the fingerprint, gel and/or CB map window (using the `Build` option) alongside clones from the displayed contig. This allows the user to determine whether they should be merged and by how much.

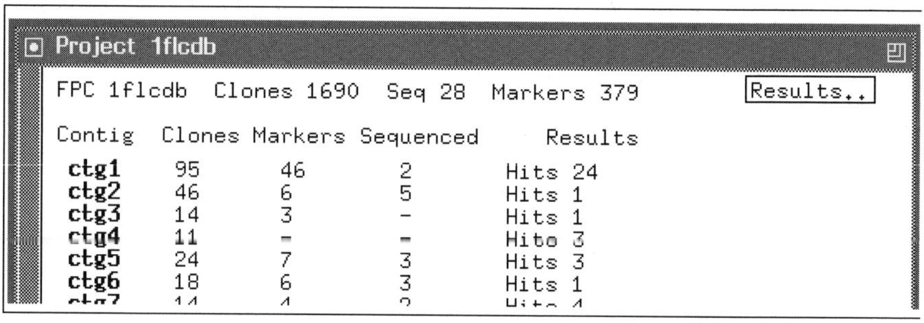

```
┌──────────────────────────────────────────────────────────────────┐
│ ▣ Project  1flcdb                                              ▥  │
│  ┌───────────────────────────────────────────────────────────┐   │
│  │ FPC 1flcdb   Clones 1690   Seq 28   Markers 379  │Results..│  │
│  │                                                   └─────────┘  │
│  │ Contig  Clones Markers Sequenced      Results               │
│  │   ctg1    95     46       2       Hits 24                   │
│  │   ctg2    46      6       5       Hits 1                    │
│  │   ctg3    14      3       -       Hits 1                    │
│  │   ctg4    11      -       -       Hits 3                    │
│  │   ctg5    24      7       3       Hits 3                    │
│  │   ctg6    18      6       3       Hits 1                    │
│  │   ctg7    14      4       2       Hits 4                    │
└──────────────────────────────────────────────────────────────────┘
```

Figure 7.12 *When* `Singles-->Ctg` *or* `Keyset-->Fpc` *is run from the* `FPC` `Analysis` *window, the project window is shown where the contigs that had 'hits' are listed at the top. The 'Hits' is the number of clones that match one or more clones in the contig.*

```
 ┌─────────────────────────────────────────────────────────────────┐
 │ ▣ Project  1flcdb                                             凹  │
 │  ┌──────────────────────────────────────────────────────────┐   │
 │  │ FPC 1flcdb   Clones 1690   Seq 28   Markers 379  │Results..││   │
 │  │                                                            │   │
 │  │ Contig  Clones Markers Sequenced      Results              │   │
 │  │   ctg2    46      6        5     FN   75  1p20  FN  17  1p18│   │
 │  │   ctg9    13      1        -     NN   65  1p18              │   │
 │  │   ctg12   11      1        -     NN    2  1p18  NN  71  1p19│   │
 │  │   ctg16    7      3        -     NN   78  1p21              │   │
 │  │   ctg17    9      2        1     FN   71  1p19  FF  24  1p22│   │
 │  │   ctg18    3      2        -     FN   33  1p18              │   │
 │  │   ctg19    5      -        -     FN   26  1p18  FN  71  1p20│   │
 │  │   ctg21    3      -        -     FN  104  1p19  FN  29  1p18│   │
 │  │   ctg22    6      1        -     FN   88  1p19              │   │
 │  └──────────────────────────────────────────────────────────┘   │
 └─────────────────────────────────────────────────────────────────┘
```

Figure 7.13 `Ends-->Ends` *results in the display of this project window. The two top-scoring candidate neighbouring contigs are listed for each. 'F' means flip, so 'FN' means the first contig needs to be flipped and the second does not. The second number is the contig (e.g. 28 in the first line is ctg28). The NpM is N = the number of clones matched, and M = the highest overlapping number of bands.*

6.3.9 Adding Markers

Markers can be added from an external file (see Section 7.2) or interactively through the marker editor. A marker can be associated with any number of clones within any number of contigs. Selecting a marker highlights all the clones which are associated with it in a contig.

The framework map markers are added from an external file as described in Section 7.2. They are displayed in the project window by selecting the `Framework` option as shown in Figure 7.14.

```
 ┌─────────────────────────────────────────────────────────────────┐
 │ ▣ Project  22cdb                                             凹  │
 │  ┌──────────────────────────────────────────────────────────┐   │
 │  │ FPC 22cdb   Clones 18122   Seq 255   Markers 896 │Framework..││ │
 │  │ 2.8  Date: 11:43 Thu 01 May 1997   User: cari              │   │
 │  │                                                            │   │
 │  │ stWI-203        509/38                                     │   │
 │  │ stD22S416       65/7 299/7 0/1                             │   │
 │  │ stD22S50.2      112/4 1180/1 175/1                         │   │
 │  │ stWI-403        112/10                                     │   │
 │  │ stF8VWFP        112/5                                      │   │
 │  │ stWI-290        112/3 7/1                                  │   │
 │  │ stIGKVP3        7/22 190/1                                 │   │
 │  │ stD22S24        7/20                                       │   │
 │  │ stD22S9         7/14 0/3                                   │   │
 │  └──────────────────────────────────────────────────────────┘   │
 └─────────────────────────────────────────────────────────────────┘
```

Figure 7.14 *Project window showing the* `Framework` *option. The markers are displayed by name in bold. The figures immediately adjacent indicate the contig number in which the marker is found and the number of clones in that contig with the marker before and after the '/' respectively.*

6.3.10   Selecting Sequence Ready Clones

The user determines sequence ready clones by viewing the amount of overlap between clones. Using the clone editor, the state is set to TILE; there are additional states to signify where a clone is in the sequencing pipeline. All the clones picked for sequencing can be highlighted in a contig.

6.3.11   Practical Tips for Selecting Sequence Ready Tile Paths

There are several criteria that should be met before a bacterial clone can be selected as being part of a contig tiling path and passed into the sequencing pipeline. The clone must have some type of unique identifier, preferably STS content data which anchors it to a long-range chromosome map, and it should be in a contig which has several-fold coverage and is of reasonable size. The STS content data serve to validate the existence of fingerprint overlap between clones and to identify possible overlaps between clones that may otherwise be lost in more tenuous matches at lower probabilities. Depth in coverage allows the 'best' clone to be selected in the tiling path based on quality of fingerprint and size of clone. Additionally, if problems are encountered with deletion in subcloning for sequencing, an alternative clone can be picked. Overlapping clones in the tiling path should share approximately 20% of their band content and the bands shared between the two clones should also be contained within another clone bridging the overlap. We routinely send our tiling path clones for fluorescence *in situ* hybridization as a means of providing confirmation of the chromosomal location.

6.3.12   Customization

Under `File...` are three functions, `Merge ace file`, `Submit for sequencing`, `Save as ace`, which are site specific. They read ACEDB file dumps from our chromosome-specific databases and write files to be loaded into the databases (see Section 7.2). Because different sites have different data representation and data flow, no effort was made to make these routines versatile. But the three functions are executed by the three C files in fpc/file called ace_merge.c, pace_save.c and ace4_save.c, which can easily be adapted to the specific sites data flow (contact cari@sanger.ac.uk for details).

7   PROCESS TRACKING AND COORDINATION

7.1   Overview

The strategy that we follow for construction of sequence ready contigs involves incorporation of experimental data from different sources in order to produce high-quality integrated maps. We believe that by exploiting all the available information in this way we can identify conflicts in the data and achieve a high level of verification. However, each experimental approach requires specialist applications environments tailored to the specific needs. For instance, construction of RH maps uses different datasets and algorithms from the assembly of contigs using restriction enzyme fingerprint data. Our approach to these conflicting requirements has been to use applications software specific to the experimental approach, and to link the applications modules through a common database format for data storage and display, and a common file format for data exchange between the modules and the database. For historical reasons our database program is ACEDB (Durbin and Thierry-Mieg, 1994; Dunham *et al.*, 1994; Dunham and Maslen, 1996) and we use the .ace text file format for data transfer (http://www.sanger.ac.uk/Software/Acedb/Acedocs). Thus we can broadly distinguish between the applications software such as Z-RHMAPPER, SAM, Image and FPC, and the data storage and display environment of ACEDB. The distinction is not absolute because each of the applications has some database functionality, and ACEDB contains some applications software.

We can also distinguish between two different modes of data transfer and update. Some data types such as STS content data for bacterial clones or the status of a shotgun sequencing project can accumulate or change on a day to day basis, and it is beneficial for all applications that need this data to be updated regularly. Other data types such as maps have changes at less frequent intervals. For instance an RH map will be recalculated only when a significant amount of new data has been obtained. In other cases where there is substantial user interaction, such as in the choice of sequence tile paths, update is required at a time specified by the user. We treat these different modes as follows.

Regular, frequent, data transfer is achieved by automated or semiautomated procedures using a range of scripts. A text mode interface called tace is available which enables a series of queries to be enacted on a specific ACEDB database (Durbin and Thierry-Mieg, 1994), and these commands can be incorporated into a shell script which is run nightly using the system cron daemon (a 'cronjob'). Data transfer that is required at irregular or infrequent intervals is initiated by manual specification within the appropriate application. In each case the data are exported from the donor software in ace format to a disk location from which it can be imported into the recipient application.

7.2   Implementation

Figure 7.2 shows the data flow between databases and applications in our sequence ready map strategy.

First, framework map coordinates for the markers are exported manually from SAM or Z-RHMAPPER to the project-specific ACEDB database as an .ace format file, where they can be displayed using the ACEDB gMap display. Each application program contains a function to produce an .ace format map file (see section 4).

Second, the STS markers are used to isolate bacterial clones and the STS content data are stored in the project ACEDB database. Meanwhile the clones that have been identified are subjected to the fingerprinting procedure and their associated bands are entered from the gel images via Image into FPC, where contigs are assembled. To merge the contigs assembled by fingerprinting with the STS content data, a nightly script extracts all the relationships between clones and STSs, and writes an .ace file of this information into a directory called 'updates' in the FPC project directory. The script specifies the target ACEDB database from which to extract the data, a series of tace commands to query the target database for clones with STS data, and some Unix shell commands to place the data into the appropriate update directory. From there the STS content data can be imported manually into FPC via the `Merge ace file` option.

Third, FPC also uses the framework map to order contigs relative to each other (Figure 7.14). The framework map data are supplied in a file called 'framework' in the FPC project updates directory which is generated manually from the ACEDB database, and is imported into FPC at the same time as the STS content data. The 'framework' file is produced by making a table using the ACEDB table maker, as shown in Figure 7.15, which is then exported as an ascii file with ';' as the separator between columns.

Fourth, sequence ready contigs can be exported manually to the ACEDB database as and when necessary, to place them in the context of all the other available information. FPC provides functions to generate .ace files of the contigs with either the minimal tiling set (`Submit for sequencing`) or the complete contig (`Save as Ace`). These files can be parsed directly into the ACEDB database either manually or with an automated tace script. Included in the contig .ace files is information to layer the positions of the contigs back on to the framework map, which can be viewed in the ACEDB gMap display (Figure 7.16).

Finally, the progress of each clone in the minimal tiling set through the shotgun sequencing process is tracked in another ACEDB-based database (locally called 'Pace'). We also keep FPC and the project database up to date with this information using a tace script in a nightly cronjob, to generate a file called sequencesnew.ace. This file contains the status tag for every clone belonging to the project and can be imported directly into FPC or ACEDB.

It should also be emphasized that, because ACEDB has its data structures specified by a set of models, the exchange of data between the databases is simple only when each database has the same model structures. See http://www.sanger.ac.uk/HGP/db_query/humace_models.shtml.

```
Quit  Help  Print  Read Def  Write Def  Pick Def
Save Def  Search Whole Class  Search Active KeySet
Create Definition  Suppress Definition

   Sort column: 1                    F4 to interrupt
   Title  Table definition for framework file
   Parameters
Column
   1  Title   STS landmark
      Width  12  Visible       Class  STS
      To restrict the search, use the condition box
      Condition

   2  Title   Project
      Width  12  Visible  Mandatory   Class  Map
      From  1     Tag:  Map
      Condition Chr_22

   3  Title   Map position
      Width  12  Visible  Mandatory   Float
      Right_of 2   Tag:  HERE # Position
      Condition
```

```
Quit  Help  Print  Switch  Export KeySet  Define Table
Map  Ascii Dump  Import Active KeySet
                                            371 lines
Table definition for framework file

STS landmarkProject        Map position

st11/60      Chr_22        4030.287109
st11/60CA    Chr_22        3986.732910
st22-5       Chr_22        23863.568359
st96T3       Chr_22        15790.132813
st031xc11    Chr_22        37831.054688
st179xa9     Chr_22        34563.000000
st234zh4     Chr_22        26795.142578
st248wd1     Chr_22        24453.199219
st268yg1     Chr_22        36703.277344
st291ve5     Chr_22        10891.644531
st320yg5     Chr_22        31486.707031
st331wc9     Chr_22        1802.228516
st443        Chr_22        3768.959961
st443CA      Chr_22        3725.406006
sta043tf9    Chr_22        18012.595703
sta048wa5    Chr_22        33249.335938
sta134yf9    Chr_22        10937.000977
sta151xe9    Chr_22        33607.476563
sta175vd5    Chr_22        24181.062500
sta190we5    Chr_22        13749.095703
sta223yd9    Chr_22        22230.738281
sta239wh1    Chr_22        12070.911133
sta247xa1    Chr_22        15108.788086
```

Figure 7.15 Screen dump of the table maker definition and the resulting table to construct the framework marker file.

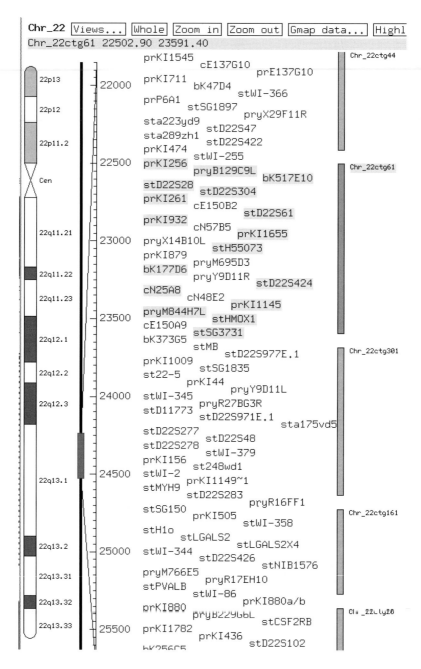

Figure 7.16 *The ACEDB gMap display showing the positions of sequence ready contigs (stippled boxes) relative to the long-range framework map. Contig Chr_22ctg61 has been picked and is highlighted. Markers that have been placed into this contig also become highlighted.*

8   CONCLUSIONS

Rapid progress in determination of the DNA sequence of the human genome depends on the availability of bacterial clones ready to sequence. We believe that this is best achieved through the rapid production of accurate bacterial clone maps, and the selection of the optimal minimal tiling set of clones from these maps. We have described a suite of software packages, which can be integrated to provide a system suitable for construction of sequence ready contigs in any genome project. Each element is freely available via the Internet and documentation is also available as indicated above.

ACKNOWLEDGEMENTS

The authors thank all the members of the Sanger Centre who have been involved with the genome sequencing projects and related software development. In particular they thank Gareth Maslen and Carol Scott for their work in integrating the system, Ian Longden for programming in FPC, Darren Platt, Friedemann Wobus and Richard Durbin for Image, and Simon Dear and Gos Micklem for their work on genomic sequencing tracking. Many thanks also go to the members of the Sanger Centre fingerprinting group for feedback and bug reports on Image and FPC, and to Panos Deloukas for his input on the RH mapping software. They also thank John Collins and Andrew Mungall for critical review of the manuscript. This work was supported by The Wellcome Trust.

REFERENCES

Collins JE, Cole CG, Smink LJ, Garrett CL, Leversha MA, Soderlund CA, Maslen GL, Everett LA, Rice CM, Coffey AJ, Gregory SG, Gwilliam R, Dunham A, Davies AF, Hassock S, Todd CM, Lehrach H, Hulsebos TJM, Weissenbach J, Morrow B, Kucherlapati RS, Wadey R, Scambler PJ, Kim U-J, Simon MI, Peyrard M, Xie Y-G, Carter NP, Durbin R, Dumanski JP, Bentley DR and Dunham I (1995) A high density YAC contig map of human chromosome 22. *Nature* **377** (**Supplement**), 367–379.

Coulson A, Sulston J, Brenner S and Karn J (1986) Toward a physical map of the genome of the nematode *Caenorhabditis elegans*. *Proc. Natl. Acad. Sci. U.S.A.* **83**, 7821–7825.

Cox DR, Burmeister M, Price ER, Kim S and Myers RM (1990) Radiation hybrid mapping: a somatic cell genetic method for constructing high-resolution maps of mammalian chromosomes. *Science* 250, 245–250.

Dunham I and Maslen GL (1996) Use of ACEDB as a database for YAC library data arrangement. *Methods Mol. Biol.* 54, 253–280.

Dunham I, Durbin R, Thierry-Mieg J and Bentley DR (1994) Physical mapping projects and ACEDB. In *Guide to Human Genome Computing*, pp. 110–158 (ed. M Bishop). Academic Press, London.

Durbin R and Thierry-Mieg J (1994) The ACEDB Genome Database. In *Computational Methods in Genome Research*, pp. 45–55 (ed. S Suhai). Plenum Press, London.

Fleischmann RD, Adams MD, White O, Clayton RA, Kirkness EF, Kerlavage AR, Bult CJ, Tomb JF, Dougherty BA, Merrick JM, McKenney K, Sutton G, FitzHugh W, Fields C, Gocayne JD, Scott J, Shirley R, Liu LI, Glodek A, Kelley JM, Weidman JF, Phillips CA, Spriggs T, Hedblom E, Cotton MD, Utterback TR, Hanna MC, Nguyen DT, Saudek DM, Brandon RC, Fine LD, Fritchman JL, Fuhrmann JL, Geoghagen NSM, Gnehm CL, McDonald LA, Small KV, Fraser CM, Smith HO and Venter JC (1995) Whole-genome random sequencing and assembly of *Haemophilus influenzae* Rd. *Science* 269, 496–512.

Green ED and Olson MV (1990) Chromosomal region of the cystic fibrosis gene in yeast artificial chromosomes: a model for human genome mapping. *Science* 250, 94–98.

Gregory SG, Soderlund C and Coulson A (1997) Contig assembly by fingerprinting. In *Genome Mapping, A Practical Approach* (ed. PH Dear), Oxford University Press, Oxford (in press).

Gyapay G, Schmitt K, Fizames C, Jones H, Vega-Czarny N, Spillett D, Muselet D, Prud'Homme JF, Dib C, Auffray C, Morissette J, Weissenbach J and Goodfellow PN (1996) A radiation hybrid map of the human genome. *Hum. Mol. Genet.* 5, 339–346.

Hudson TJ, Stein LD, Gerety SS, Ma J, Castle AB, Silva J, Slonim DK, Baptista R, Kruglyak L, Xu S-H, Hu X, Colbert AME, Rosenberg C, Reeve-Daly MP, Rozen S, Hui L, Wu X, Vestergaard C, Wilson KM, Bae JS, Maitra S, Ganiatsas S, Evans CA, DeAngelis MM, Ingalls KA, Nahf RW, Horton LT, Anderson MO, Collymore AJ, Ye W, Kouyoumjian V, Zemsteva IS, Tam J, Devine R, Courtney DF, Renaud MT, Nguyen H, O'Connor TJ, Fizames C, Faure S, Gyapay G, Dib C, Morissette J, Orlin JB, Birren BW, Goodman N, Weissenbach J, Hawkins TL, Foote S, Page DC and Lander ES (1995) An STS-based map of the human genome. *Science* 270, 1945–1954.

Ioannou PA, Amemiya CT, Garnes J, Kroisel PM, Shizuya H, Chen C, Batzer MA and de Jong PJ (1994) A new bacteriophage P1-derived vector for the propagation of large human DNA fragments. *Nature Genet.* 6, 84–89.

Kim U-J, Shizuya H, Kang H-L, Choi S-S, Garrett CL, Smink LJ, Birren BW, Korenberg JR, Dunham I and Simon MI (1996) A bacterial artificial chromosome-based framework contig map of human chromosome 22q. *Proc. Natl. Acad. Sci. U.S.A.* 93, 6297 6301.

Riles L, Dutchik JE, Baktha A, McCauley BK, Thayer EC, Leckie MP, Braden VV, Depke JE and Olson MV (1993) Physical maps of the six smallest chromosomes of *Saccharomyces cerevisiae* at a resolution of 2.6 kilobase pairs. *Genetics* 134, 81–150.

Riley J, Butler R, Ogilvie D, Finniear R, Jenner D, Powell S, Anand R, Smith JC and Markham AF (1990) A novel, rapid method for the isolation of terminal sequences from yeast artificial chromosome (YAC) clones. *Nucleic Acids Res.* 18, 2887–2890.

Roach JC, Boysen C, Wang K and Hood L (1995) Pairwise end sequencing: a unified approach to genomic mapping and sequencing. *Genomics* **26**, 345–353.

Schuler GD, Boguski MS, Stewart EA, Stein LD, Gyapay G, Rice K, White RE, Rodriguez-Tome P, Aggarwal A, Bajorek E, Bentolila S, Birren BB, Butler A, Castle AB, Chiannilkulchai N, Chu A, Clee C, Cowles S, Day PJR, Dibling T, Drouot N, Dunham I, Duprat S, East C, Edwards C, Fan JB, Fang N, Fizames C, Garett C, Green L, Hadley D, Harris M, Harrison P, Brady S, Hicks A, Holloway E, Hui L, Hussain S, Louis-Dit-Sully C, Ma J, MacGilvery A, Mader C, Maratukulam A, Matise TC, McKusick KB, Morissette J, Mungall A, Muselet D, Nusbaum HC, Page DC, Peck A, Perkins S, Piercy M, Qin F, Quackenbush J, Ranby S, Reif T, Rozen S, Sanders C, She X, Silva J, Slonim DK, Soderlund C, Sun WL, Tabar P, Thangarajah T, Vega-Czarny N, Vollrath D, Voyticky S, Wilmer T, Wu X, Adams MD, Auffray C, Walter NAR, Brandon R, Dehejia A, Goodfellow PN, Houlgatte R, Hudson JR, Ide SE, Iorio KR, Lee WY, Seki N, Nagase T, Ishikawa K, Nomura N, Phillips C, Polymeropoulos MH, Sandusky M, Schmitt K, Berry R, Swanson K, Torres R, Venter JC, Sikela JM, Beckmann JS, Weissenbach J, Myers RM, Cox DR, James MR, Bentley D, Deloukas P, Lander ES and Hudson TJ (1996) A gene map of the human genome. *Science* **274**, 540–546.

Shizuya H, Birren B, Kim UJ, Mancino V, Slepak T, Tachiiri Y and Simon M (1992) Cloning and stable maintenance of 300-kilobase-pair fragments of human DNA in *Escherichia coli* using an F-factor-based vector. *Proc. Natl. Acad. Sci. U.S.A.* **89**, 8794–8797.

Slonim D, Kruglyak L, Stein L and Lander E (1996) Building human genome maps with radiation hybrids. RECOMB 97, ACM. *Proceedings of the First Annual International Conference on Computational Molecular Biology*, pp. 277–286, sponsored by ACM DIGACT, ACM Press.

Soderlund C (1995) *SAM: User's Manual*. Technical Report SC-01-95. The Sanger Centre, Hinxton Hall, Cambridge, UK.

Soderlund C and Dunham I (1995) SAM: a system for iteratively building marker maps. *CABIOS* **11**, 645–655.

Soderlund C and Longden I (1996) *FPC V2.5: User's Manual*. Technical Report SC-01-96. The Sanger Centre, Hinxton Hall, Cambridge, UK.

Soderlund C, Longden I and Mott R (1997) FPC: a system for building contigs from restriction fingerprinted clones. *CABIOS* **13**(s) (in press).

Stein L, Kruglyak L, Slonim D and Lander E (1995) RHMAPPER software package. http://www-genome.wi.mit.edu/ftp/pub/software/rhmapper.

Sulston J, Mallett F, Staden R, Durbin R, Horsnell T and Coulson A (1988) Software for genome mapping by fingerprinting techniques. *CABIOS* **4**, 125–132.

Venter JC, Smith HO and Hood L (1996) A new strategy for genome sequencing. *Nature* **381**, 364–366.

Wu C, Zhu S, Simpson S and de Jong PJ (1996) DOP-vector PCR: a method for rapid isolation and sequencing of insert termini from PAC clones. *Nucleic Acids Res.* **24**, 2614–2615.

8 Software for Human Genome Sequencing

Simon Dear

The Sanger Centre, Wellcome Trust Genome Campus, Hinxton, Cambridge CB10 1SA,
UK E-mail: sd@sanger.ac.uk

1 INTRODUCTION

The accurate determination of the sequence of nucleotide bases in a genomic region involves many steps, only the last of which is DNA sequencing. Usually the region of interest is isolated to one or more clones, whose vector inserts are found to overlap during the process of physical mapping. A minimally overlapping set of clones is selected for sequencing to minimize redundancy. Current sequencing methods rely on the electrophoretic separation of their reaction products, the resolution of which is considerably less than the typical size of the clone being sequenced. Therefore, the clone is invariably fragmented into shorter pieces, which are themselves cloned and later sequenced. The biology of subcloning and sequencing is not addressed here. The process of reconstruction of the sequence of a clone from the sequences of its many subclones is described in the context of large-scale sequencing of the human genome. The key informatics problems will be discussed with reference to the software that has been developed to address them.

There are as many approaches to the informatics of large-scale genome sequencing as there are laboratories involved in sequencing. There can be no prescriptive solution because there is no single software package that addresses all issues. Nor are there widely agreed standards used by software developers that would allow their programs to be used interchangeably. However, almost all programs that are suitable for large scale sequencing have been developed for Unix and are readily available. An investigator undertaking genome sequencing must bring together programs developed by many authors, and often has to tune them to his or her particular needs. Although this approach involves a significant investment of resources, the advantage is that the investigator is able to adapt the system to their preferred strategy and to new problems as they arise.

2   OVERVIEW

The 'shotgun' sequencing strategy has proven itself to be a robust and efficient method for accurately determining sequences of anything from a few kilobases to hundreds of kilobases in size. This approach involves the generation of a random set of subclones made from the parent clone by physical shearing or enzymatic digestion. The location and orientation of these subclones is not known *a priori*. However, if two subclones overlap, this can be detected by comparing their sequences (or 'reading') obtained from sequencing. Thus their precise relationship can be determined and, if enough subclones are sequenced, the sequence of the clone itself can be reconstructed. This reconstruction process is called 'sequence assembly'. A sequence or partial sequence determined in this way is often called a 'consensus'.

The number of subclones required for sequence assembly depends on the size of the clone and the number of base pairs that can be reliably determined when a subclone is sequenced. A 100-kb P1 artificial chromosome (PAC) may require a 'shotgun' of 1800 subclones.

After sequence assembly it may be found that the whole clone is not represented. Instead there may be several sets of contiguous readings (or 'contigs') with gaps of unknown size in between. The expected number of contigs can been determined theoretically (Lander and Waterman, 1988) or by simulation, using a program such as RANCL (J. Sulston, unpublished results). In practice, the precise size of a clone is often not known, the readings may vary in quality and the subcloning is not random. Therefore, a common strategy is to start with a fixed number of readings based on a fixed shotgun redundancy (six fold, say), and, if the assembly proves fragmentary, additional shotgun readings are made.

The size of the shotgun is a trade-off between cost and the number and quality of the contigs obtained. At some point, however, it will be cost effective to switch to a directed strategy. In a directed phase, individual subclones will be resequenced using different sequencing methods, with the aim of bridging contig gaps and improving contig quality in areas where it is poor. This directed phase is called 'finishing' and is performed by a person called a 'finisher'.

3   GEL IMAGE PROCESSING

Commercial automated gel sequencing machines, such as the ABI 377, Pharmacia A.L.F. and the LiCor 4000L, have dedicated microcomputers and proprietary software for data capture and gel image analysis. Typically, a composite image is constructed during the sequencing gel run and afterwards is analysed, producing for each loaded sample a chromatogram representing the profile of the lane in which it ran, and its sequence. These data are then transferred to Unix workstations where preprocessing and sequence assembly will be performed.

Gel image processing involves several algorithmic steps. These are normally applied in the order listed, although certain dye and gel technologies may make some steps unnecessary. Some Unix-based programs exist.

3.1   Lane Tracking

Several samples are loaded into wells at the top of the gel. During electrophoresis, the samples migrate through the gel and appear as lanes on the gel image. The gel image is analysed to identify each lane. Automated approaches, such as GETLANES (Cooper *et al.*, 1996), have met with varied success. In practice, each lane must be checked manually and edited if found incorrect.

3.2   Trace Extraction

A profile of each lane is generated. This chromatogram contains four time-series traces, one for each base. In systems where the laser beam scans backwards and forwards across the gel, a time correction is applied.

Commercial systems generate proprietary trace formats. However, SCF (Dear and Staden, 1992) has been adopted widely as a standard by independent software developers.

3.3   Colour Deconvolution

In multiple fluorescent dye systems, the emission spectra of the dyes can and often do overlap. Colour deconvolution determines the contribution of each labelled base, from known emission properties of each dye.

3.4   Background Subtraction

There is excitation even on an empty gel, and this background signal is removed from each trace.

3.5   Mobility Correction

Fragment mobility depends on its length and on the properties of the fluor-escent dyes. Differing dye mobilities cause the traces for each dye to become out of register with one another.

3.6   Peak Normalization and Smoothing

Peak heights are standardized and traces are smoothed by removing high-frequency components.

3.7   Base Calling and Quality Estimation

Peaks are identified, and their order determines the sequence. Estimations of base calling accuracy can be calculated.

PHRED (P. Green and B. Ewing, personal communication) is a widely used base calling package. It analyses processed traces in ABI or SCF formats, and produces base calls and associated quality values. Much effort has been made to calibrate the quality values to make them meaningful. On the basis of the trace characteristics, PHRED computes a probability p of an error in the base call at each position, and converts this to a quality value q using the transformation $q = -10 \log_{10} p$. Thus a quality of 30 corresponds to an error probability of 1/1000.

These values are clearly useful in estimating the quality of individual read-ings and have been used effectively by sequence assemble programs, such as PHRAP and for contig sequence generation (Bonfield and Staden, 1995).

3.8   Unix-Based Gel Image Processing Packages

Several Unix-based packages and components have been developed for gel image processing, mostly by large genome centres with the aim of streamlining their operations and improving the quality of their sequence data.

GELMINDER (D. Platt, unpublished results) was designed to manage gel image processing in high-throughput laboratories, and is used routinely in the Sanger Centre, Hinxton, and the Genome Sequencing Center, Saint Louis, Missouri. It incorporates GETLANES, several lane editors, PLAN (B. Ewing, personal communication) for trace processing and PHRED. The package has dramatically reduced the processing time of each gel, and has improved sequence quality.

GRACE (L. Stein, unpublished results) is a graphical interface to the BASS program which incorporates lane tracking, trace extraction and base calling algorithms. BASS derives heavily from SAX (Berno, 1996). GRACE produces SCF trace files, sequence and quality files compatible with PHRED, although the quality measures themselves are not directly comparable owing to insufficient statistical calibration.

3.9   Data Transfer

Depending on whether proprietary or Unix-based software is to be used for gel analysis, trace and sequence data or gel images will need to be transferred to Unix for further processing. Mechanisms for transfer will depend on the local network that is available and the type of microcomputer used by the sequencing machine. Without being specific, some observations are made.

It is most likely that the microcomputer will be connected to the Unix system via Ethernet. This is satisfactory for the transfer of trace files (which can each be 100 kbytes). Transfer of a gel image (40 Mbytes or more) will take considerable time and create significant network load.

Data for ABI 377s are collected on Apple Macintosh computers. The packages Ethershare and Columbia Appletalk Protocol both implement the Appletalk network protocol over Ethernet, and are satisfactory for small volumes of data. For large volumes, these are inefficient, and FTP should be considered. Alternative network protocols, such as Asynchronous Transfer Mode (ATM), should be considered if extremely large data volumes are required to be transferred routinely. We have observed that a 40-Mb gel image can take over 15 min to transfer via Ethershare, only 10 min when using FTP, and under 1 min across an ATM network.

4   SEQUENCE PREPROCESSING

Sequences obtained from fluorescence-based sequencing machines consist of a mixture of good and unreliable data, and may contain fragments of unwanted sequence such as vector and library contaminants. Although some sequence assembly programs make allowances for these (PHRAP is a notable example), detection and removal before assembly avoid unnecessary recomputation later, and provide valuable feedback on sequence and library quality. Preprocessing should be restricted to identifying features particular to individual readings or subclones rather than those of the clone being sequenced, for the simple reason that the latter is more reliably determined from the consensus generated during assembly.

4.1   Quality Clipping

Many sequence assembly programs assume that the sequences presented to them have a low error rate, and therefore require that poor-quality regions are clipped from readings before assembly. PHRAP does not require clipped data. Average reading length statistics generated during clipping provide a useful and objective measure of sequence quality.

The program TREV (Figure 8.1), a rewrite of the trace viewing program TED (Gleeson and Hillier, 1991), allows the user to remove poor data from the start and end of each reading. This is a labour-intensive and highly subjective procedure. However, when readings are a part of the directed phase and have been selected to resolve specific problems in the assembly, the manual approach is appropriate.

A common automated approach is to look for clusters of unresolved bases (Ns) within the sequence, indicative of a degradation in quality, and to clip just before this (Bonfield and Staden, 1996). The success of this method depends on the base calling software producing Ns, and in our experience works well for ABI data but less so for PHRED. Also the presence of sequence compressions can result in premature clipping.

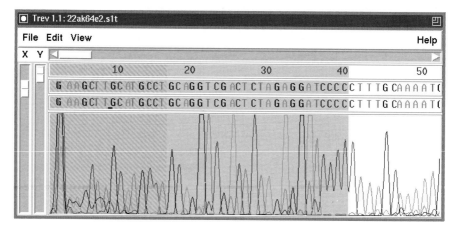

Figure 8.1   *The TREV trace editor in operation. The first 50 bases of a trace are shown. The greyed out portion of the trace has been clipped because it is vector. Quality clipping is performed in an independent operation, and the results of both are recorded in experiment file output.*

An alternative method is to estimate the signal to noise ratio in the trace, and to clip when its value, over an appropriately chosen window, reaches an unacceptable level. Later versions of TED have incorporated this feature.

If PHRED quality values are available, effective clipping can be performed by identifying the highest scoring window of bases after a fixed value (we suggest 15) has been subtracted from each quality value.

4.2   Sequencing Vector Identification

The first step in the reconstruction of the sequence of the clone is to identify and remove all sequencing vector, because it is a product of subcloning.

Sequencing vector upstream of the cloning site is found at the start of the reading. If the insert of the subclone is shorter than the length of the reading, vector sequence may also be present at the end.

It is important to have the vector sequence and knowledge of the location of cloning sites and sequencing primers before attempting to remove vector, as these priors greatly assist correct detection, especially in poor-quality regions.

Several approaches to automated vector clipping have been developed. The program VECTOR_CLIP (written by R. Staden) uses hashing and treats the problem like a dot matrix comparison. This approach is fast, and works well with vector rearrangements.

When the vector lies in low-quality data, its presence may be obscured. Dynamic programming, although slower, is more sensitive in these cases. CROSS_MATCH (P. Green, personal communication) is a fast implementation of the Smith–Waterman alignment algorithm and has an option '-screen' for masking regions matching vector, or any other target sequence.

Vector at the start of the reading is often difficult to identify because the primer peak saturates the trace signal. SVEC_CLIP (R. Mott, unpublished results) aligns the trace to the expected vector sequence. This method can identify vector when there is little or no information in the base calls themselves.

Often, several methods must be tried, because no single method works in all cases.

4.3   Cloning Vector Identification

Cloning vector will not be present in regions already identified as sequencing vector, and, if it is present at all, it will comprise one end or all of the reading. Both VECTOR_CLIP and CROSS_MATCH perform this function well.

It is important to know the vector sequence and the location of the cloning site. VECTOR_CLIP requires that the cloning site be located at the start of the sequence file.

Cloning vector is more easily identified by comparing the vector sequence with the consensus generated during sequence assembly. However, early identification provides useful feedback on library quality. As an example, an unexpectedly high presence of cloning vector may indicate clone deletion.

4.4   Removal of Contaminants

Clones grown in bacterial hosts are likely to have low levels of contamination from the host genome. In particular, yeast artificial chromosome (YAC) clones are highly susceptible to contamination with yeast. Additionally, individual libraries may be infected with phage, such as the bacteriophage N4.

This task is made easier with the availability of the full genomes of *Escherichia coli* and *Saccharomyces cerevisiae*. A quick and effective way to screen for these contaminants is to compare the part of the reading not already identified as vector against a database of the host genome using the program BLASTN (Altschul *et al.*, 1990).

To minimize the rate of false-positive matches, we suggest using the BLASTN parameters 'N = −4, X = 3, W = 18, S = 250'. With these, BLASTN will report only exact matches of at least 50 base pairs.

4.5   Unix-Based Preprocessing Packages

Most of the programs described above are non-interactive and are best run from a script, which applies the necessary processing steps to all readings. Several software packages have been developed for this purpose. PREGAP (Bonfield and Staden, 1996), a part of the Staden package, can take a batch of data from a variety of sequencing machines, gather information required for processing the reading, identify the good quality data, and mark sequencing and cloning vector and Alu repeats. ASP (D. Hodgson, unpublished results) performs similar tasks and has a graphical interface which allows the user to select the readings to be processed and the processing steps to use. After processing, it gathers statistics which form the basis of detailed status reports.

4.6   Experiment Files

Both PREGAP and ASP use Experiment Files (Bonfield and Staden, 1996) as a

Table 8.1   Descriptions of commonly used experiment file record types

Code	Description
ID	IDentifier
SF	Sequencing vector File
SC	Sequencing vector Cloning site
SP	Sequencing vector Primer site (relative to cloning site)
SL	Sequencing vector present at Left end of reading
SR	Sequencing vector present at Right end of reading
QL	poor Quality sequence present at Left end of reading
QR	poor Quality sequence present at Right end of reading
SQ	the reading's SeQuence
CC	a Comment
TG	a TaG to be placed on a reading
TC	a Tag to be placed on the Consensus

Mnemonics are shown by capital letters.

mechanism for exchanging information between processing modules. Experiment Files are designed to be easy to write and parse. They are textual. Each record starts with a two-letter code. (Table 8.1). The file must start with an ID record. The remaining records can appear in any order.

An initial Experiment File is created and contains all the information necessary to process each reading. This includes the location of sequence vector files, details of cloning sites and primer locations, the name of the associated trace file and the sequence of the reading itself. Each processing step reads the information it requires, performs its computation, then adds its results to the end of the file. The sequence itself is not modified; rather, it is annotated.

The Staden package web site (http://www.mrc-lmb.cam.ac.uk:80/pubseq/) has a full description of the format.

5   SEQUENCE ASSEMBLY

The process of sequence assembly reconstructs the sequence of a clone from readings made from shorter DNA fragments generated from the clone.

The Staden assembly program GAP4 (Bonfield *et al.*, 1995) constructs the contigs by incorporating one reading at a time. The reading is compared with the current consensus to find its best location. If the reading aligns sufficiently well, it is added to the contig. Contigs are joined if the reading spans a gap. The consensus is then recalculated. These steps are repeated for all remaining readings. Changing the order in which readings are presented to the program can result in a different assembly.

GAP4 also has a directed assembly mode, where prior knowledge of the reading's location can be used to constrain its placement. This information is conveyed in the experiment file. This is applicable to the assembly of finishing readings, which should lie close to other readings made from the same subclone.

The program GAP4 is considerably more than a sequence assembly engine. It incorporates many facilities for the editing and finishing. In contrast, other programs such as PHRAP and FAKII (Larson *et al.*, 1996) are purely sequence 'assembly engines'.

Sequence assembly in GAP4 is not restricted to the internal algorithm. Interfaces exist to external engines, such as FAKII and CAPII (Huang, 1992). Hopefully new engines will be incorporated as they become available. Preassembled readings can be imported into GAP4 using the 'Enter preassembled data' option, provided overlap information has been included in the experiment files.

PHRAP exhaustively compares all readings against all in a pairwise fashion, ranking each potential reading pair overlap according to its alignment score. Starting with the highest scoring pair, it then takes each pair in turn and constructs contigs by greedy algorithm. CAPII functions in a similar manner. Graph theoretical approaches have been used, such as in FAKII.

5.1   Assembly Pragmatics

What distinguishes these assembly programs in practice is the way they address the experimental nature of the data, and the presence of sequence artifacts.

Sequences from gel readings tend to be reliable in the first few hundred bases, with a rapid degradation of quality towards the end of the gel run. GAP4 requires that readings are clipped conservatively, although unused data are visible in the GAP4 editor, and can later be incorporated if required. PHRAP uses the full length of the reading and the extra sequence helps it to find the correct location of the reading and allows it to make joins between contigs that would otherwise be missed.

Sequence artifacts, such as compressions, unclipped sequencing vector, and chimera or pseudo-chimera, can also disrupt assembly. These often fail to assemble into the right place, if at all, and have to be either added or removed manually in what is normally a tedious process. PHRAP has many pragmatic features for dealing with these. If pairs of readings are found to match at their starts then diverge thereafter, unclipped vector is assumed.

Repetitive DNA poses significant problems during assembly. If the size of the repeat sequence is similar to or greater than the length of a typical reading, misassembly (the wrong placement of a reading) can occur. Repetitive regions often have to be reassembled at higher stringencies, possibly over several iterations. PHRAP uses base quality information to adjust its pairwise alignment score when there are discrepant high-quality bases between pairs of readings. This makes the pair of readings less likely to be assembled together. When used with the CONSED editor (D. Gordon, unpublished results), which allows the user to set the quality of both discrepant bases to an arbitrarily high value, this is a powerful method for resolving misassembled regions.

The Alu repeat is about 300 bases long, comparable to the good quality part of a reading. As a result, GAP4 has difficulty assembling Alu-rich clones, and strategies have been proposed to deal with this. These normally involve assembling non-Alu readings first, then assembling readings containing Alu afterwards at a higher stringency. PHRAP uses the full length of the reading, which greatly diminishes the problem.

The sequence for a contig is derived from the sequence of its aligned readings. This is often calculated by taking the consensus of the individual aligned readings at each base position. This may be either a majority vote or a weighted measure based on individual base qualities (Bonfield and Staden, 1995). PHRAP uses PHRED quality values. The consensus is taken from the reading with the highest quality base at each position. This 'golden path' approach is able to skirt around localized problems if there are good-quality readings present. PHRAP is able to assign a quality value to each base in the consensus This value is derived from the quality of the base used after it has been moderated by the quality of supporting and conflicting bases. The value is considered to be a conservative representation of the true accuracy of the base.

5.2   Assembly Interchange Formats

A standard interchange format for sequence assembly is not available, although several have been proposed. GAP4 uses the Experiment File format, with extensions to record the location of readings within the contig. The Whitehead Institute Genome Center has proposed the Boulder data exchange format.

We have promoted our own Common Assembly Format (CAF). CAF is based on the textual data files used by the ACEDB (Durbin and Thierry-Mieg, 1997) and is directly compatible with the ACEMBLY assembly environment (J. Thierry-Mieg, unpublished results). This format is easily amenable to manipulation by scripts. Efficient tools have been developed to interface with PHRAP and FAKII (Myers, 1996), and for the creation and dumping of GAP4 databases.

CAF has allowed us to develop modular processing software. Cloning vector removal is performed automatically directly after assembly. An autoeditor resolves base-pair discrepancies between individual readings and the consensus, by reference to aligned readings and their traces. Finishing readings in the directed phase are selected automatically by the program FINISH (G. Marth and S. Dear, unpublished results). These programs are encapsulated in the PHRAP2GAP package.

6   EDITING AND FINISHING

After sequence assembly, additional reading and editing is usually required to raise the accuracy of the sequence to a level that is considered finished. This process is called finishing, and involves the design and assembly of additional sequencing experiments, and the visual inspection of assembly discrepancies. Table 8.2. shows approximate error rates at different stages of sequencing.

The target of one error in 10 000 base pairs is widely accepted as necessary for reliable gene recognition and primer design. It is also essential that the final edited consensus be assembled correctly, and is a true representation of the clone sequence. It is extremely hard to estimate the accuracy of a consensus, although PHRAP quality measures are helpful. In practice, finishing criteria are set and must be met before a sequence is considered finished. Although these do not guarantee that the quality target is reached, they greatly increase confidence in the final sequence.

Finishing criteria are at the discretion of the investigator, but several are commonly held. The sequence should be contiguous; that is, it should be without gaps. There should be no unresolved bases (Ns). Each base should be represented by at least two subclones. This minimizes errors resulting from single base-pair polymorphisms in individual subclones. Each base should be represented by readings of complementary strands or different sequencing chemistries. This minimizes undetected compressions or base drop-outs during sequencing. When there are direct tandem repeats of substantial number, they should be quantified by independent experiment, such as sizing a PCR fragment containing the repeat.

Table 8.2   Redundancy during sequencing reduces the error rate in contigs over unedited readings. Finishing improves the quality further

Sequence	Error rate
Raw reading	More than 1 in 100
Assembled consensus	About 1 in 1000
Finished sequence	Less than 1 in 10 000

Editing and finishing are the processes through which these quality criteria are met. Editing involves identifying significant discrepancies in the assembly, and resolving them using human judgement. This should be minimized owing to its subjective nature, although for many problems there are sufficient data to resolve them without recourse to additional experimentation, especially if the expertise of the trained finisher is utilized. When there are insufficient data, additional readings should be obtained to lend support to the existing data.

6.1   Editing in GAP4

GAP4 is a widely used program for sequence editing. The program is a complete sequence assembly and finishing tool, and has many sophisticated features for managing large-scale assembly projects. Its efficient, well designed interface is a significant advancement on that of earlier versions of the program (Dear and Staden, 1991).

GAP4 stores readings and overlap information in a database of a custom format. Readings are imported and exported as Experiment Files. Consensus sequences can be generated in FASTA and other formats.

All functions are initiated from a main window. This window also shows all output messages generated during the editing session in a scrollable window. A second window, the Contig Selector, shows the relative sizes of each contig and the location of all tags. The Contig Selector also doubles as a two-dimensional (2D) plot showing matches from many of the internal search routines ('Find Repeats', 'Find Read Pairs', 'Find Internal Joins', 'Find Oligos' and 'Check Assembly').

An important feature is the Contig Editor (Figure 8.2). This shows the alignment of readings within a narrow window (80 bp) of the contig. The consensus is shown below the readings. Scroll-bars and buttons aid navigation through the contig. Traces from which the readings were derived are viewable in a separate window. It is through reviewing the evidence in the traces that discrepancies between readings and the consensus are resolved.

Editing proceeds by finding the next place in the contig where the consensus is unresolved. The 'Next Problem' button relocates the cursor at this position. After reviewing the trace data, it may be decided that there are sufficient data to resolve consensus ambiguity by editing, or that there are readings assembled into the wrong location. Bases in readings can be edited either by overtyping with the correct base, by adjusting quality values, or by adjusting the clipping to exclude the base from inclusion in the consensus. Wrongly assembled readings can be disassembled using 'Remove Reading' for later reassembly back into the database. It is also possible to break the contig into two pieces, if gross misassembly is found.

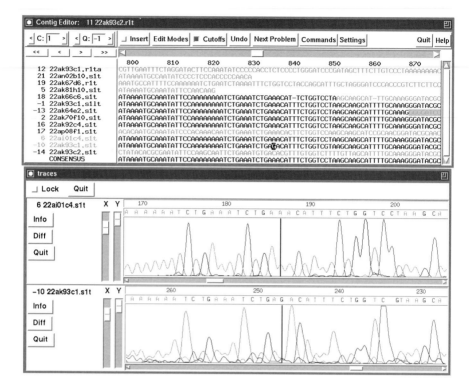

Figure 8.2   *The GAP4 Contig Editor (top) and Traces windows (bottom). Aligned readings are shown. Bases of clipped data appear as grey. Traces, through which discrepancies are resolved, are viewed by double clicking on the base of interest.*

Contig overlaps missed during sequence assembly can be found using 'Find Internal Joins'. This option shows candidate overlaps on the 2D plot, and each must to be reviewed manually in the Contig-Joining Editor before committing the join.

The editing facilities inside GAP4 are extremely powerful, although the process is highly labour intensive, subjective, and not automatically repeatable except by repeating the steps manually.

6.1.1   Designing Finishing Readings

When there are insufficient data in a region, (i.e. the finishing criteria are not met), additional readings are required. Before designing new sequencing experiment, it is prudent to check that no existing yet unassembled readings are helpful. The 'Screen only' mode of the Assembly option can be used to find readings that fall into problem areas.

Additional readings are normally obtained by resequencing selected subclones using a different method, as sequencing artifacts are often sequencing chemistry or strand specific. For example, dye terminator chemistry could be used to complement existing dye primer readings. Custom oligo primers can be designed using the 'Select primer' option (Hillier and Green, 1991), if no subclone ends lie near the problem.

Contig gaps completely lack any data, and must be filled by designing primers near contig ends and walking off a suitable subclone, or by resequencing subclones using a method that yields a longer sequence.

The optimal finishing strategy depends on the expectation of success and accuracy of the reading required, and the cost of the reagents. These are likely to vary between laboratories, and the finishing strategy therefore should be decided on locally.

6.1.2   Assembling Readings

GAP4 includes a sequence assembler that will take readings, in Experiment File format, and add them to existing contigs. This allows finishing readings to be incorporated efficiently into edited contigs without reassembling all the data.

6.2   Editing in CONSED

CONSED (Figure 8.3), an editor for PHRAP assembled data, is becoming used more widely. CONSED does not have a database of its own, but takes reading information from the PHD files generated by PHRED, and overlap information from the file created by PHRAP when the '-ace' option is used.

The editor, although superficially like GAP4's Contig Editor, reflects a different editing and finishing philosophy. Problems are found by searching for low consensus quality or for high-quality discrepancies. Low-quality regions require additional data to bring them up to the target quality value of 40, which corresponds to less than one error per 10 000 bases. Any manual operation that cannot be reproduced by reassembly with PHRAP is not supported. For example, it is not possible to remove individual readings, nor to break or join contigs. Readings can be edited, although this is discouraged in preference for editing base qualities. In areas of misassembly, discrepant bases can be marked high quality so that they are less likely to be placed together after reassembly with PHRAP. After several interations of editing and reassembly, many problems concerning tandem repeats can be addressed.

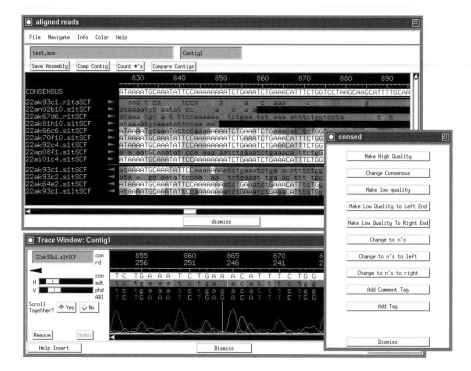

Figure 8.3   *The CONSED editor (top) and trace window (bottom), along with the menu of base editing commands (right). The background colour of each base is coloured in a grey-level to indicate its quality. The region shows poor reading support on the bottom strand.*

6.3   Automated Finishing

The process of the design of resequencing experiments for finishing can be automated. There are two steps. The first, is to identify regions that require additional data. This can be rule based, such as finding areas that fail to meet finishing criteria, or based on directly addressing low-quality areas, such as those with low PHRAP consensus quality. The second step is to produce a list of subclones and sequencing methods that, when resequenced, will address the problems identified in the first step. Producing an optimal list is computationally hard, and heuristics, such as greedy algorithms, are often used. Automatic selection of custom primers should be made with reference to quality information to ensure the sequence is reliable, as oligo synthesis is costly.

GAP4 has several automated features. 'Suggest Long Reads' and 'Suggest Primers' identify finishing readings that resolve regions with single strand and chemistry. 'Suggest Probes' identifies custom primers that lie on ends of contigs, for walking or contig gap-filling experiments. 'Compressions and Stops' chooses terminator chemistry readings for areas tagged as such.

FINISH (G. Marth and S. Dear, unpublished results) uses overlap and chemistry information contained in a CAF file to determine regions not meeting finishing criteria. The set of sequencing methods that are able to resolve each problem type can be specified. FINISH uses a greedy algorithm to identify the theoretical reading which maximizes the expectation of resolving the problem and minimizes the cost. Its contribution is evaluated, using reading length distribution profiles. If the problem is still not satisfactorily resolved, the next most useful reading is considered. The proposed list is an efficient, although suboptimal, set of readings to perform.

Automated finishing can be performed just after shotgun assembly. When sequenced, the finishing readings can then be incorporated into the assembly. Several iterations of this can be performed before the clone is given to the finisher for manual editing and finishing.

7   FEATURE IDENTIFICATION

Sequence features within the clone can aid assembly, suggesting both contig orientation and ordering, and also places where misassembly may have occurred. During editing, the finisher can be directed to areas where finishing is not required.

7.1   Repeat Sequences

All repetitive elements are a possible cause of misassembly. Therefore it is important to identify them and to confirm the assembly across the regions. Human repeats families are well described. Alu repeats occur frequently in the human genome. Misassembly is possible but unlikely owing to frequent base-pair differences between Alu family members. Other known repeats cause problems infrequently because they are rarer, and are unlikely to occur twice in the clone.

Large exact and near-exact repeats invariably cause significant problems. There are two classes of repeats: inverted and directed. Inverted repeats are either dispersed or have hairpin structures. The presence of large dispersed inverted repeats may go unnoticed by the finisher, and the orientation of the intervening sequence should be confirmed. Arrays of direct repeats are notoriously hard to assemble when their combined length is greater than the insert of the subclone. Differences between each repeat element are often sufficient for the region to be assembled through painstaking human effort. It is not always possible to determine precisely the number of copies of the repeat. Cases of variable number of copies of tandem repeats have been observed. Their number can be estimated experimentally, such as by sizing a PCR fragment that spans the region.

Many programs are suitable for detecting repeats. REPEATMASKER (A. Smit and P. Green, unpublished results) identifies and characterizes human repeat families. Hidden Markov Models, such as implemented in the HMMER package (Eddy, 1996) provide a sensitive statistical technique for sequence matching, and have been applied successfully to the identification of Alu family members (G. Micklem, personal communication). BLASTN and CROSS_MATCH also work well.

7.2   When Not to Finish

It is unlikely that the whole of the sequence need be finished and edited. Cloning vector sequence should be identified and not edited.

If many clones are being sequenced from the same region of a physical map, effort may be duplicated if the full insert of each clone is finished. This can be avoided by finishing the overlap in only one clone. When following this strategy, it is important to allow a small overlap, say of 100 base pairs, to aid the piecing together of the genomic region. Overlaps can be detected after assembly by comparing the consensus sequences of both clones using a program such as CROSS_MATCH. Clones may be polymorphic in the overlap region, and these variations should be annotated.

E. coli transposons, such as Tn1000 (accession no. X60200 in the EMBL Data Library), may be present in the host strain and may have contaminated the clone. The sequence for the insertion need not be finished, and should be removed from the final sequence. It is possible for there to be a mixed population of subclones, some with the transposon and some without. Sequencing a transposon of around 6000 bp long represents a considerable overhead, and where possible transposon-free host strains should be used.

7.3   Contig Ordering

Large-scale features, such as LINE1 repeats, transposons, cloning vector and overlaps with physical map neighbours, can be used to orient and order contigs if the feature spans contig gaps. CROSS_MATCH is a suitable program for identifying large-scale features for which the sequence is known.

Contigs may also be ordered when readings from the same subclone made with different sequencing primers fall in different contigs. It has been observed that a significant proportion of readings are misnamed owing to lane drop-outs undetected during gel image processing. Therefore, spanning read pairs should not be taken as hard evidence for joins unless several pairs span the same gap in a consistent fashion.

7.4   Tagging and Viewing Features

The programs described above for feature identification function on a consensus sequence in FASTA format and produce text output. However, the results of the analysis may be viewed in the context of the aligned readings.

Both GAP4 and CONSED allow features to be tagged for viewing within their contig editors. Tags are also visible in several GAP4 windows, including the Contig Selector and Template displays.

Tags can be imported into GAP4 with the 'Enter Tags' option. This reads descriptions of the new tags from a file, and adds them to aligned readings (TG records) or to the consensus (TC records). The format is a subset of an Experiment File, and close enough for Experiment Files to be used directly. The following is an example of a definition of a cloning vector tag to be placed on the consensus of a contig containing the named reading. The third line is a comment attached to the tag.

```
ID   6aas73e5.s1
TC   CVEC + 1721..2215
TC        CLONING VECTOR: PAC 16026 15530 16026
```

The following defines a warning tag to be placed on a named reading:

```
ID   6abg76h6.q1eU
TG   WARN + 570..702
TG        POSSIBLY VECTOR: pUC18 251 387 2686
```

CAF can also be used to create tags within a GAP4 database. The CAF system comes with a Perl language toolkit and subroutines for adding tags to both readings and consensus. A CAF file can be generated from a GAP4 database, be modified to include new tags, and then used to create a new annotated GAP4 database.

The consensus can also be analysed in graphical viewers. SIP (Staden, 1982) can show the similarity between two sequences in a dot-plot display. The program DOTTER (Sonnhammer and Durbin, 1996) shows a dot-plot and allows zooming, dynamic adjustment of the match sensitivity, and display of local ungapped alignments about the cursor position. PRINTREPEATS (Parsons, 1995) produces a graphic showing regions of similarity. The result of BLASTN searches can be browsed in BLIXEM (Sonnhammer and Durbin, 1994). ACEDB has been used to view the results from many analysis programs.

8   QUALITY CONTROL

Once the clone has been finished to the level where it meets the finishing criteria, it is important to apply independent quality control procedures, if only to confirm confidence in the sequence.

The most fundamental procedure is to have another person visually inspect the assembly in a contig editor to ensure that the minimum quality criteria are met in all places. Graphical viewers that show outstanding problem areas, such as HAWK (D. Hodgson, unpublished results), can aid this process.

GAP4 has several functions to aid checking. The Quality Plot shows places of insufficient sequence accuracy and strand coverage, and its use after finishing can confirm that no region has been overlooked. It is also sensible to check regions where extensive editing has taken place. The Contig Editor can search for bases in the consensus for which there was no evidence in the original readings. If it is common practice to enter all manually edited bases in lowercase letters, 'Highlight Disagreements' will indicate all places where bases were changed. Clusters of edits should be checked carefully.

Restriction digests can be used to detect misassembly. The clone is fully digested with a restriction enzyme and the resulting fragments are run on a gel with a size standard. An image of the gel is taken and the sizes of the fragments, which appear as bands on the gel, are estimated and compared with the theoretical digest of the clone. Theoretical digests can be calculated using the program NIP (Staden, 1994). In practice, the procedure is repeated with different enzymes to maximize the chance of misassembly detection.

Pairs of readings made with forward and reverse sequencing primers from the same subclone can confirm the assembly. If the pair is consistently located and oriented, confidence in the intervening sequence is increased. The program CROW (C. Brown, personal communication) presents a histogram of pairs spanning each base. Dips in the plot correspond to areas where support is poor, and may indicate misassembly.

Another independent check is to reassemble all unedited readings with a different assembly program, as with the PCOP procedure (L. Hillier, unpublished results). Discrepancies between the consensus sequences should be reported. If the assembly program assigns quality values to the consensus, high-quality discrepancies should be checked.

As mentioned above, inverted repeats are often a source of misassembly. Judicious use of sequence self-comparison tools like DOTTER can alert their presence to the checker.

9   THE FINISHED SEQUENCE

The end result of the sequencing process is the production of high-quality sequence. This sequence is then analysed using comparative and predictive methods to identify genes and other features. These are annotated, and submitted with the sequence to a public sequence data repository. Coverage of these procedures is beyond the scope of this chapter.

However, it is important to point out that the output of the sequencing process may be more than just the sequence. All places where sequence accuracy may be called into question should be annotated clearly.

If a region was found to contain a tandem repeat with many copies, it may not be possible to determine the exact number. Further, repeats may cause clone instability, resulting in deletion. Mixed population of subclones, each with a different number of repeats, may be observed and should be reported.

Some hairpin loops and poly-GC tracts are resilient to subcloning and may cause contig gaps. Alternative sequencing strategies should be tried and, if after reasonable effort none yields results, the gap should be sized. If there are more than two contigs, the order and orientation of internal contigs should be determined.

The location of *E. coli* transposons, which should be excised from the final sequence, should be recorded.

Polymorphisms found by comparison with clones previously sequenced should be annotated as variations, as they may be artifacts of cloning or sequencing.

The location of the start and end of overlapping clones should be annotated if their position is known.

It is important to remember that we are sequencing a DNA fragment as it is represented in the clone, and not the genome. The presence of cloning artifacts, such as variable numbers of tandem repeats and single base pair mutations, is inevitable from time to time.

Currently there is no agreed quantitative description of sequence accuracy. Although PHRAP will produce per base consensus quality measures, these values are not available after editing. A qualitative description consisting of observations of the above nature is at least a move in the direction of providing useful information about sequence accuracy.

10   CONCLUSION

This chapter has been intended as an overview of the central issues in the informatics of human genome sequencing. Many individual software packages have been described or referred to. While not attempting to be prescriptive, it has also not been possible exhaustively to review all the available programs. The novice user should use this chapter as a road map, not as a shopping list, and is encouraged to explore the many highways and byways that lead from it. Some packages, such as the Staden package, offer near-complete solutions for small sequencing projects.

Much of what has been described is directly applicable to genome projects of organisms other than human which are being undertaken by sequencing of mapped clones. Many of the programs can handle large insert clones efficiently, such as YACs, PACs and BACs. Shotguns of whole microbial genomes or chromosomes require special treatment.

Sequencing software is driven by the need to assemble larger and larger shotguns, and to produce finished sequence more quickly and accurately. The subjective elements of editing need to be reduced greatly, leaving the finisher to tackle the difficult problems. There is much left to do.

11 FINDING SOFTWARE ON THE WEB

Information about all the software described in this chapter can be found on the World Wide Web, or by mailing the author or contact person. Some software packages are available by anonymous file transfer protocol (FTP). Please consult authors regarding licensing policy.

ACEDB, ACEMBLY, ASP, BLIXEM, CAF, CROW, DOTTER, GELMINDER, HAWK, PHRAP2GAP and SVEC_CLIP
E-mail: Simon Dear, sd@sanger.ac.uk (general queries)
URL: http://www.sanger.ac.uk/Software/

BASS and GRACE
URL: http://www-genome.wi.mit.edu/ftp/pub/software/Bass/

BLASTN
URL: http://www.ncbi.nlm.nih.gov/Home.html

CAP (Columbia Appletalk Protocol)
URL: http://www.cs.mu.oz.au/appletalk/cap.html

CAPII (Contig Assembly Program)
E-mail: Xiaoqiu Huang, huang@mtu.edu
URL: ftp://ftp.bio.indiana.edu/molbio/align/

CONSED
Email: David Gordon, gordon@genome.washington.edu (academic), Gerald Barnett, barnett@u.washington.edu (commercial)
URL: http://www.genome.washington.edu/consed/consed.html

CROSS_MATCH, PHRAP and PHRED
E-mail: Phil Green, phg@u.washington.edu
URL: http://bozeman.genome.washington.edu/phrap_documentation.html

FAKII. Fragment Assembly Kernel
URL: http://www.cs.arizona.edu/research/reports.html
E-mail: Gene Myers, gene@cs.arizona.edu

GAP4, NIP, PREGAP, SIP, TED, TREV and VECTOR_CLIP (The Staden Package)
E-mail: Rodger Staden, rs@mrc-lmb.cam.ac.uk
URL: http://www.mrc-lmb.cam.ac.uk/pubseq/

GETLANES lane tracking software
URL: http://genome.wustl.edu/gsc/gschmpg.html

HMMER (Hidden Markov Model package)
E-mail: Sean Eddy, eddy@genetics.wustl.edu
URL: http://genome.wustl.edu/eddy/HMMER/main.html

OSP (Oligo Selection Program, whose search engine is used in GAP4)
E-mail: LaDeana Hillier, lhillier@watson.wustl.edu

PLAN
E-mail: Brent Ewing, bge@u.washington.edu

REFERENCES

Altschul SF, Gish W, Miller W, Myers EW and Lipman DJ (1990) Basic local alignment search tool. *J. Mol. Biol.* **215**, 403–410.

Berno AJ (1996) A graph theoretic approach to the analysis of DNA sequencing data. *Genome Res.* **6**, 80–91.

Bonfield JK and Staden R (1995) The application of numerical estimates of base calling accuracy to DNA sequencing projects. *Nucleic Acids Res.* **23**, 1406–1410.

Bonfield JK and Staden R (1996) Experiment files and their application during large-scale sequencing projects. *DNA Seq.* **6**, 109–117.

Bonfield JK, Smith KF and Staden R (1995) A new DNA sequence assembly program. *Nucleic Acids Res.* **23**, 4992–4999.

Cooper ML, Maffitt DR, Parsons JD, Hillier L and States D (1996) Lane tracking software for four-color fluorescence-based electrophoretic gel images. *Genome Res.* **6**, 1110–1117.

Dear S and Staden R (1991) A sequence assembly and editing program for efficient management of large projects. *Nucleic Acids Res.* **19**, 3907–3911.

Dear S and Staden R (1992) A standard file format for data from DNA sequencing instruments. *DNA Seq.* **3**, 107–110.

Durbin R and Thierry-Mieg J (1997) http://www.sanger.ac.uk/Software/Acedb.

Eddy S (1996) Hidden Marker Models. *Curr. Opin. Struct. Biol.* **6**, 361–365.

Gleeson T and Hillier L (1991) A trace display and editing program for data from fluorescence based sequencing machines. *Nucleic Acids Res.* **19**, 6481–6483.

Hillier L and Green P (1991) OSP: an oligonucleotide selection program. *PCR Methods and Applications* **1**, 124–128.

Huang X (1992) A contig assembly program based on sensitive detection of fragment overlaps. *Genomics* **14**, 18–25.

Lander ES and Waterman MS (1988) Genomic mapping by fingerprinting random clones: a mathematical analysis. *Genomics* **2**, 231–239.

Larson S, Jain M, Anson E and Myers E (1996) *An Interface for a Fragment Assembly Kernel*. Technical Report TR96-04. Department of Computer Science, University of Arizona, Tucson, AZ.

Myers E (1996) *A Suite of UNIX Filters for Fragment Assembly*. Technical Report TR96-07. Department of Computer Science, University of Arizona, Tucson, AZ.

Parsons J (1995) Miropeats: graphical DNA sequence comparisons. *Comput. Appl. Biosci.* **11** 615–619.

Sonnhammer ELL and Durbin R (1994) A workbench for large scale sequence homology analysis. *Comput. Appl. Biosci.* **10** 301–307.

Sonnhammer ELL and Durbin R (1996) A dot-matrix program with dynamic threshold control suited for genomic DNA and protein sequence analysis. *Gene* **167** GC1–10.

Staden R (1982) An interactive graphics program for comparing and aligning nucleic acid and amino acid sequences. *Nucleic Acids Res.* **10**, 2951–2961.

Staden R (1994) The Staden Package. In *Methods in Molecular Biology* (eds AM Griffin and HG Griffin), Vol. 25, Humana Press, Totawa, NJ, pp. 9–170.

9   Human EST Sequences

Guy St C. Slater

UK Human Genome Mapping Project Resource Centre, Hinxton, Cambridge CB10 1SB, UK

1   INTRODUCTION

Expressed sequence tags (ESTs), the short segments produced by partial sequencing of cDNA clones, are being generated rapidly in huge quantities; they now outnumber all other publicly available sequence and thus provide a considerable information resource. However, these data are not ideal: the sequences are incomplete, they contain a high frequency of errors, and they are generated in highly redundant sets.

This chapter covers the various publicly available sources of EST data, derived 'added value' databases, and the tools available for their analysis.

2   BACKGROUND

Gene density in the human genome is exceptionally low; it is estimated that only about 3% is actually coding. The bulk of useful information about the human genome can be gained from the coding regions; these allow elucidation of the corresponding protein sequences. In addition, these sequences can be obtained in a fraction of the time required for whole genome sequencing. It is due to both of these factors that justification is found for giving priority to sequencing the expressed genome over a total genome sequencing strategy.

An EST is defined (Strachan and Read, 1996) as being 'A short sequence of a cDNA clone, for which a PCR assay is available'. Although primer pairs have not been made for most sequences in the EST databases to attempt to amplify genomic DNA, these complementary DNA (cDNA) fragments are now commonly referred to as ESTs.

Gerhold and Caskey (1996) explain that 'Looking at all available 5' and 3' ESTs and then trying to infer a gene's function is like reading a random chapter and epilogue from a book and then guessing at the plot'.

The information yielded from cDNA sequencing is far from complete. To extend the analogy further, the book shop that is the human genome contains are many copies of popular novels, while several rare books are unavailable.

3   GENERATION OF THE EST DATA

3.1   Making the cDNA Libraries

Before ESTs can be generated, a cDNA library must first be made. To do this, messenger RNA (mRNA) is extracted from the tissue of interest, from which cDNA is made using reverse transcriptase. The cDNA is then isolated and made into a double-stranded copy using DNA polymerase. Short linker sequences containing a restriction site are ligated on to the blunt ends of this cDNA. The cDNA is then digested so that it will contain overhanging ends complementary to those on the vector into which is it ligated. The rest of the procedure is similar to conventional directional cloning: these recombinant molecules are transformed into competent cells; the product of one gene on the vector will allow selection of cells that contain the vector; and the absence of the product of another gene which spans the insertion point will allow selection of cells containing the insert.

More cDNAs will be generated for strongly expressed genes (as there will be more copies of the mRNA present), thus methods are designed to try to reduce the number of cDNAs present from a single gene as this will reduce the redundancy in the resultant dataset. Two methods are commonly used to reduce the redundancy caused by strongly expressed genes. Subtractive hybridization is used to group together and eliminate cDNAs that already appear in another library, and cDNA library normalization is a technique in which incomplete annealing of single-stranded sequences will cause the more prevalent cDNAs to anneal. These can then be removed, leaving a library with a reduced degree of redundancy.

3.2   Generating the ESTs

Primers complementary to the sequence flanking the cloning (or insertion) site on the vector are used to initiate random sequencing from both the 5' and 3' ends. These are described as being *universal* primers, as they are specific to the vector and not the insertion sequence. About 200–500 nucleotides are sequenced in a single pass from both the 5′ and 3′ ends, but the 3′ EST often contains only the sequence of the untranslated region at the end of the mRNA.

3.3 Mapping

Physical maps have been constructed using STSs (Hudson *et al.*, 1995) to provide the scaffolding for large-scale genome sequencing. Similarly, once the redundancy in the data has been removed, EST-based physical maps can be constructed Aaronson *et al.*, 1995; Houlgatte *et al.*, 1995; Korenberg *et al.*, 1995; Schuler *et al.*, 1996). These provide the groundwork for large-scale sequencing of coding sequences. ESTs can play an important role in the mapping of a genome; once a gene has been tagged, the site can be placed on a physical map. As each EST points to an expressed gene, these maps are thus *expression* or *transcript maps* (Boguski and Schuler, 1995).

The mapping of the cDNAs is essential for identification of the chromosomal locations of disease genes represented within EST datasets, and ESTs can now be assigned to a specific chromosome band rapidly and accurately using (fluorescence *in situ* hybridization (FISH)) or radiation hybrid mapping.

4 REASONS FOR EST SEQUENCING

4.1 Advantages

Initially, much of the interest in EST sequencing was in anticipation of the possibility of allowing patenting of the ESTs as molecular probes. This is still a highly controversial issue, as such a patent could imply rights to the full-length gene which it encodes.

This approach to sequencing can be used to generate a huge number of sequences in a short amount of time. As can be seen in Figure 9.1, ESTs have been generated at such a rate that they now outnumber all other sequences. The EST database from the NCBI, dbEST, (Boguski *et al.*, 1993), now contains more than a million ESTs.

As the sequences generated are derived from mRNA, they will have already undergone splicing to remove introns, and thus may be translated in an attempt to obtain a partial sequence of the corresponding protein.

Often the reverse transcription utilized during the preparation of the cDNA libraries is incomplete. This results in different parts of the gene being covered

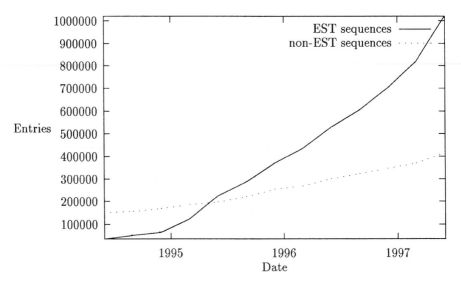

Figure 9.1 *ESTs now outnumber all other sequences in the EMBL database (Stoesser et al., 1997).*

by the 5' sequencing from different cDNA clones. Thus the sequence data can be treated as though they were data from a shotgun sequencing project, and the ESTs assembled to generate a consensus for a much longer region of the gene than that which would be covered by a single EST.

4.2   Disadvantages

Although the ESTs are long enough to allow homology searching against a database of well characterized proteins, they are normally too short to cover the entire gene, and so the information will be incomplete until data have been assembled derived from many cDNAs associated with a single gene; this is unlikely to occur with the weakly expressed genes.

ESTs are also very error prone as they are generated by single-pass sequencing; error rates of 2–5% have been reported. The error rate is also likely to increase along the length of the EST; the further the sequencing reaction has progressed from the primer, the higher the error rate.

There can also be contamination due to portions of the vector having been sequenced. In addition to this, the presence of low complexity and repetitive sequence such as poly(A) sequences and Alu repeats can hinder the analysis of the data. Sequencing of non-nuclear-derived mRNAs can also occur (i.e. mitochrondrial or ribosomal RNA derived sequences).

Another danger is the presence of chimeric cDNA clones, whereby the vector contains multiple inserts (derived from different loci). This is caused by coligation of multiple fragments, or by recombination between multiple recombinant DNA molecules in a single cell.

Although there are currently over 780 000 human ESTs in dbEST, it is estimated that there only 50 000 to 100 000 different human genes; thus there is a huge degree of redundancy in the data. For example, in the UniGene dataset, the α-chain haemoglobin cluster (strongly expressed) currently contains 3277 ESTs. This is due to the different rates at which genes are expressed within various tissues; however, assumptions cannot be made about expression patterns when redundancy reduction methods have been used such as subtractive hybridization and cDNA library normalization.

5 PUBLICLY AVAILABLE EST DATA

5.1 Raw Data

There is a huge amount of publicly available EST data. Three principal sources are the EMBL (Stoesser *et al.*, 1997), GenBank (Benson *et al.*, 1997) and DDBJ, Tateno and Gojobori (1997). Data are exchanged between these three on a daily basis, so the content of these databases is essentially synchronized. Each of these databases has a division comprised entirely of EST sequences. Also available is dbEST (Boguski *et al.*, 1993), a database specifically for EST data; this contains the same sequences as the GenBank EST division, but offers more detailed annotation.

5.2 Analysed Data

There are a number of 'added value' EST databases available, all sharing the aim of generating a non-redundant gene index in which information for each gene is stored in a discrete group.

The majority of these databases have been generated by the use of similar strategies. First, quality control is applied to the raw data (e.g. Adams *et al.* (1995) discarded all ESTs composed of more than 3% N or shorter than 100 bp). Then the sequences are clustered into groups, normally employing an heuristic word-based algorithm for the sequence comparison; this clustering is usually performed using the 3' ESTs, with the 5' ESTs being assigned to the cluster of their 3' counterparts, thus promoting an organization in which each cluster represents all the data for a single gene.

Clustering of the data also allows the best representative subset of clones to be found so that a non-redundant cDNA library can be designed for hybridization experiments. In addition, it facilitates the identification of novel gene families and enables effective cross-species comparison studies to be performed between clustered EST sets from different organisms.

Many of these databases contain a consensus sequence for each cluster which has been generated by assembly of the cluster members. This is considerably more useful than a simple cluster of sequences. For example, a sequence comparison-based database search (e.g. BLAST), performed on a database of consensus sequences, will exhibit a better signal to noise ratio than a search performed on the corresponding database of raw EST data; the amount of data has been reduced without loss of useful information.

5.2.1   *UniGene and UniEST*

The aim of the UniGene project (Boguski, 1995; Schuler *et al.*, 1996) is to work towards the organization of human genes on a physical map – a *human transcript map*. UniGene is a non-redundant set of genes derived from the GenBank database (Benson *et al.*, 1997), whereas UniEST is a non-redundant set of ESTs that do not match any of the UniGene clusters. Release 96.0 of UniEST contains 53 356 clusters obtained from 325 488 human EST sequences.

As the datasets have been generated for physical mapping, the clustering was performed according only to the 3′ ends of the ESTs. This is because the 3′ untranslated regions (UTRs) contain fewer introns than their 5′ counterparts; thus the PCR products during mapping are unlikely to become prohibitively large owing to the presence of an intron on the genomic DNA, and because they are less conserved than the coding regions, so different genes with similar products are less likely to be clustered together.

To reduce the time taken for the clustering, an heuristic approach is used in the generation of the UniEST set; a word-based heuristic is then used to identify candidate members of a cluster, after which an alignment is generated between regions of the sequences containing matching words.

5.2.2   *Merck Gene Index*

The Merck Gene Index project (Aaronson *et al.*, 1995) is the product of a collaboration for large-scale EST sequencing. The aim is to produce a similar gene index from the large number of ESTs being generated, but it is currently unavailable.

The insert size (cDNA length) that has been cloned is also available for the Merck data; this is useful in assembling the ESTs into consensus sequences, as it allows estimation of relative positions of contributing ESTs and the sizes of any gaps.

5.2.3 TIGR HGI

The Institute for Genomic Research (TIGR) Human Gene Index (HGI) integrates TIGR's own data with a variety of other publicly available data, aiming to provide a non-redundant database, eventually representing all human genes.

Data are available on the tissue from which the contributing ESTs were derived; additional information is being added about the expression patterns, cellular roles, functions of the genes and their products, and evolutionary relationships between them. The database will ultimately include links to corresponding genomic sequences and protein structures, as well as mapping data and relevant literature.

The HGI contains Tentative Human Consensus (THC) sequences, the product of assembly of the EST data using TIGR assembler (also comprising cDNA sequence data obtained by conventional methods, where available) described as 'virtual transcripts'.

There is also cross-referencing between HGI and the TIGR EGAD (Expressed Gene Anatomy Database), a non-redundant set of mature human transcripts, derived from complete cDNA sequences or from genomic data where cDNA data are unavailable.

5.2.4 STACK and SaniGene

Recently, the STACK (Sequence Tag Alignment Consensus Knowledge-base) database has been released (currently unpublished). This is an expressed sequence consensus database, built from publicly available data. The STACK protocols are designed to maximize the consensus sequences which are built from the ESTs, and employ improved error detection methods. This is achieved by inclusion of poorer quality data which are discarded in other clustering schemes, but still considering the quality of the data while building the consensus sequences.

6 EST ANALYSIS TOOLS

6.1 ICAtools

The ICAtools package (Parsons *et al.*, 1992; Parsons, 1995) is a set of tools for clustering cDNA sequences. It uses a non-hierarchical clustering algorithm, whereby each sequence may appear in more than one cluster. Comparisons are performed using a hash table-based algorithm similar to that of Fasta (Lipman and Pearson, 1985; Pearson and Lipman, 1988). The clustering is highly dependent on the input order of the sequences, and although an n^2 comparison is avoided (where every sequence is compared against every other sequence), the clustering times reported would make this approach impractical for the extremely large datasets that are now becoming available.

6.2   TIGR Assembler

The TIGR assembler (Sutton *et al.*, 1995) has been used for EST analysis (Adams *et al.*, 1995). It is an algorithm for the assembly of the large number of sequences generated by shotgun sequencing projects; thus, when applied to EST data, the analysis is being conducted as though the ESTs were the product of a shotgun sequencing project specifically on the expressed portions of the human genome.

The TIGR assembler greatly reduces the time required for sequence assembly by rapidly identifying candidate pairs of sequences using a rapid word-based algorithm, thus determining cluster members, before using a greedy dynamic programming based algorithm to align and assemble sequences with significant overlap. The advantage of this assembly algorithm compared with its predecessors is that it can overcome the problems associated with the presence of repeat regions. The detection of repeats is based on the distribution of the number of potential overlaps for each fragment, with non-repeat regions being assembled first.

When applied to EST data, 340 000 ESTs were clustered to generate 40 000 THC sequences (and 80 000 singletons) (White and Kerlavage, 1996).

7   BIOLOGICAL RESOURCES

7.1   IMAGE

The IMAGE (Integrated Molecular Analysis of Genomes and their Expression) Consortium was established to share gridded arrays of cDNA clones and to make derived sequence, mapping and expression data publicly available. The majority of the libraries are from human tissues, but mouse cDNA libraries are also being arrayed. A main aim of the consortium is to allow reorganization of the cDNAs to develop a non-redundant master array, containing a representative cDNA from each gene in the genome.

8   WWW RESOURCES FOR ESTs

Database

EMBL	http://www.ebi.ac.uk/ebi_docs /embl_db/ebi/topembl.html
GenBank	http://www.ncbi.nlm.nih.gov/Web/Genbank/
DDBJ	http://www.ddbj.nig.ac.jp/
dbEST	http://www.ncbi.nlm.nih.gov/dbEST/

Resources

IMAGE	http://www-bio.llnl.gov/bbrp/image/image.html
STACK	http://www.sanbi.ac.za/stack/
UniGene	http://www.ncbi.nlm.nih.gov/UniGene/
TIGR HGI	http://www.tigr.org/tdb/hgi/hgi.html
Merck Gene Index	http://www.merck.com/mrl /merck_gene_index.2.html

Analysis

TIGR Assembler	ftp://ftp.tigr.org/pub/software/TIGR_assembler/
ICAtools	http://sunny.ebi.ac.uk/~jparsons/packages /icatools/ICAtools_index.html

9   SUMMARY

The vast quantity of EST data currently available in public databases provides a huge information resource. However, as EST data are inherently redundant and error prone, the information is of limited value in its raw form and, when considering a specific biological problem, a more informed hypothesis can be formulated when using a derived higher-order database.

REFERENCES

Aaronson JS, Eckman B, Blevins RA, Borkowski JA, Myerson J, Imran S and Elliston KO (1996) Toward the development of a gene index to the human genome: an assesssment of the nature of high-throughput EST sequence data. *Genome Res.* 6, 829–845.
Adams MD, Kerlavage AR, Fleischmann RD, Fuldner RA, Bult CJ, Lee NH, Kirkness EF, Weinstock KG, Gocayne JD, White O *et al.* (1995) Initial assessment of human

gene diversity and expression patterns based upon 83 million nucleotides of cDNA sequence. *Nature* 377 (**Supplement**), 3–17.

Benson DA, Boguski MS, Lipman DJ and Ostell J (1977) GenBank. *Nucleic Acids Res.* 25, 1–6.

Boguski MS (1995) The turning point in Genome research. *Trends Biol. Sci.* 20, 295–296.

Boguski MS, Lowe TMJ and Tolstoshev CM (1993) dbEST – database for 'expressed sequence tags'. *Nature Genet.* 4, 322–333.

Boguski MS and Schuler GD (1995) ESTablishing a human transcipt map. *Nature Genet.* 10, 369–371.

Gerhold D and Caskey CT (1996) It's the genes! EST access to human gene content. *BioEssays* 18, 973–981.

Houlgatte R, Mariage-Samson R, Duprat S, Tessier A, Bentolila S, Lamy B and Auffray C (1995) The Genexpress Index: a resource for gene discovery and the genic map of the human genome. *Genome Res.* 5, 272–304.

Hudson TJ, Stein LD, Gerety SS, Ma J, Castle AB, Silva J, Slonim DK, Baptista R, Kruglyak L and Xu S-H *et al.* (1995) An STS-based map of the human genome. *Science* 270, 1945–1954.

Korenberg JR, Chen X-N, Adams MD and Venter JC (1995) Toward a cDNA map of the human genome. *Genomics* 29, 364–370.

Lipman DJ and Pearson WR (1985) Rapid and sensitive protein similarity searches. *Science* 277, 1435–1441.

Parsons J (1995) Improved tools for DNA comparison and clustering. *CABIOS* 11, 603–613.

Parsons J, Brenner S and Bishop M (1992) Clustering cDNA Sequences. *CABIOS* 8, 461–466.

Pearson WR and Lipman DJ (1988) Improved tools for biological sequence comparison. *Proc. Natl. Acad. Sci. U.S.A.* 85, 2444–2448.

Schuler GD, Boguski MS, Stewart EA, Stein LD, Gyapay G, Rice K, White RE, Rodrigues-Tome P, Aggarwal A and Barjorek E *et al.* (1996) A gene map of the human genome. *Science* 274, 540–546.

Stoesser G, Sterk P, Tuli MA, Stoehr PJ and Cameron GN (1997) The EMBL nucleotide sequence database. *Nucleic Acids Res.* 25, 7–13.

Strachan T and Read AP (1996) *Human Molecular Genetics*. BIOS Scientific, Oxford.

Sutton GG, White O, Adams MD and Kerlavage AR (1995) TIGR assembler: a new tool for assembling large shotgun sequencing projects. *Genome Sci. Technol.* 1, 9–19.

Tateno Y and Gojobori T (1997) DNA Data Bank of Japan in the age of information biology. *Nucleic Acids Res.* 25, 14–17.

White O and Kerlavage AR (1996) TDB: new databases for biological discovery. *Methods Enzymol.* 266, 27–40.

10 Prediction of Human Gene Structure

Luciano Milanesi[1] and Igor B. Rogozin[1,2]

[1] Istituto Tecnologie Biomediche Avanzate, CNR, Via Fratelli Cervi 93, 20090 Segrate, Milano, Italy and [2] Institute of Cytology and Genetics SD RAS, Lavrentyeva av. 10, Novosibirsk 630090, Russia
E-mail: milanesi@itba.mi.cnr.it

1 INTRODUCTION

Gene identification is an important problem in current molecular biology studies. Due to recent progress in large-scale sequencing projects, gene identification programs have become widely used. The use of these programs can significantly simplify the analysis of newly sequenced DNA especially when applied in combination with experimental methods. The gene identification procedure is very complex owing to the structure of eukaryotic genes. Although good results have been obtained with a variety of computational approaches, the problem of gene structure prediction has not yet been solved completely.

A large number of theoretical methods have been developed for sequence analysis and prediction. More detailed descriptions of the different approaches for sequence analysis are described in the following review articles (Staden, 1984, 1985; Stormo, 1988; Fickett and Tung, 1992; Milanesi *et al.*, 1994; Fickett, 1995, 1996a; Gelfand, 1995; Snyder and Stormo, 1995a; Burset and Guigo, 1996; Fickett and Guigo, 1996). Reference databases can be found at the following sites: http://linkage.rockefeller.edu/wli/gene/; http://expasy.hcuge.ch/sprot/seqanalr.html; http://www-hto.usc.edu/software/procrustes/fans_ref.

The analysis of human genes cannot be considered merely as a linguistic analysis of the nucleotide string because the gene structure is made up of many other important features. These include higher-order chromatin structure, the non-random nucleosome positioning along the DNA, the different features of the three-dimensional structure of the DNA (or RNA) and the torsional strain on the DNA induced by transcription.

In this chapter the most important aspects of gene structure prediction are described: functional sites in nucleotide sequence; functional regions in nucleotide sequences; protein-coding gene structure prediction; analysis of potential proteins coded by predicted genes; RNA-coding gene structure prediction. The methods described in this chapter are all publicly available. A list of all internet addresses (WRL) cited in this chapter can be found at http://www.itbd.mi.cnr.it/genlinks.

2   GENOME STRUCTURE

A protein-coding gene consists of a coding sequence usually interrupted by non-coding sequences called introns. Introns can also interrupt non-coding regions (introns in 5' and 3' untranslated regions of pre-mRNA). Mammalian genes are the combination of a set of short exons dispersed along very variable lengths of intron DNA, and the distance between genes is often much larger than the genes themselves. The term 'exon' is normally applied to regions that are not spliced out from a pre-mRNA sequence (5' untranslated region (5' UTR), coding sequences (CDS) and 3' untranslated region (3' UTR)), but this term is often used also to indicate the protein coding regions only. The gene identification tools make use of both meanings.

The functional unit of a genome is a functional site (also called functional sequence, motif, signal and pattern). A combination of functional sites (often binding sites for other molecules) constitutes a functional region. Different functional sites can overlap and/or interact. The functional unit of a protein coding sequence is a codon. Codons are an example of non-overlapping encoding for a single protein. Some other functional sites can be found in protein coding sequences. The codon is a simple functional site which interacts with tRNAs. Other functional units such as protein binding sites are much more complicated.

The density of protein and RNA coding sequences in the human genome is very low. However, the non-coding part of the human genome is far from being junk DNA. The non-coding DNA is usually responsible for the complex regulation of the genome and functioning of genes. Moreover, non-coding regions can have several different functional interpretations (Trifonov, 1996).

The primary structure of the genome is very heterogeneous. The most well known example of such heterogeneity is CpG islands. CpG islands constitute a specific fraction of the genome because, unlike bulk DNA, they are non-methylated and they contain CpG at high frequencies. The CpG islands have a significantly higher G + C content than the rest of the DNA. CpG islands may be used as gene markers. There are about 45 000 CpG islands per haploid human genome (Antequera and Bird, 1993). Significant compositional differences between cold-blooded and warm-blooded vertebrates are found at the whole genome (DNA) level, as well as during analysis of homologous genes. There is a trend toward the formation of a larger component of GC-rich isochores in genomes of warm-blooded vertebrates (Bernardi, 1989; Bernardi and Bernardi, 1991). Differences in the damage and repair rates among different segments of chromatin DNA have been observed. As a result, the rates of synonymous substitutions along chromatin fibres can also vary significantly from about two to about nine substitutions per nucleotide per 10^9 years (Li and Graur, 1991; Boulinkas, 1992). Distinctive sequence features in protein coding, genic non-coding and intergenic human DNA have been observed (Guigo and Fickett, 1995) and different types of inhomogeneities have also been revealed in yeast chromosome III (Karlin *et al.*, 1993). Due to these properties the structure of the human genome is very heterogeneous and difficult to analyse by computational methods.

3 APPROACHES TO EVALUATE PREDICTION ACCURACY

Evaluating the accuracy of a computational tool is an important issue. Experiments that are planned on the basis of computer prediction often need a considerable amount of effort and resources. Accuracy is also important for making a comparison between programs developed for similar purposes. The problem of the performance evaluation of the tools used in functional sequence (site and region) prediction is a recurrent theme in computational biology. Functional sequences analysis is generally based on discrimination between two sets: a set of functional sequences and a set of sequences that do not perform this function (non-functional sequences). From the methodological point of view, the evaluation of an algorithm is a difficult task. One should split initial datasets into independent representative training and control sets. Control datasets are used only for the evaluation of accuracy.

In practice, this problem is more complicated. The sequence databases are usually highly redundant and some sequences are repeated many times. The composition of the control set must take this fact into account in order to avoid strong cross-correlation between training and control sets due to a significant portion of the similar sequences present in both the sets. The control sets are often small (at least for gene identification programs) and consist of well behaved sequences (Burset and Guigo, 1996). The criteria for sequence selection are not always well defined. Some programs have been updated regularly with new data and, due to this procedure, it is possible to obtain different results by analysing the same sequence at different times. Therefore, it is not easy to understand the real accuracy of these programs.

All the above-mentioned problems are multiplied when different programs are compared. For example, the evaluation of several gene identification programs on carefully selected control sets has shown that the prediction accuracy is systematically lower than originally announced (Burset and Guigo, 1996).

One possible way to avoid these problems is to implement standard performance metrics and datasets. Fickett and Tung (1992) developed a standardized benchmark to evaluate coding measures. Burset and Guigo (1996) carried out exhaustive tests of gene identification programs and suggested several standardized measures of accuracy evaluation. Kulp *et al.* (1996) suggested the use of both training and control datasets in order to compare gene identification methods. Lopez *et al.* (1994) used very large DNA genomic sequences to evaluate the GRAIL2 program.

Let TP be the number of correctly predicted functional sequences, TN the number of correctly predicted non-functional sequences, FN the number of incorrectly predicted functional sequences and FP the number of incorrectly predicted non-functional sequences. All methods can be characterized by the probability of underpredicting the real functional sequences (first-type error E1) or overpredicting them (portion of non-functional sequences predicted as real functional; second-type error E2). Widely used measures are:

$$Sn = 1 - E1 = TP/(TP + FN) \qquad \text{Sensitivity}$$
$$Sp = 1 - E2 = TN/(TN + FP) \qquad \text{Specificity}$$

In some cases the measure $Sen2 = TP/(TP + FP)$ is used (Guigo *et al.*, 1992; Snyder and Stormo, 1993; Dong and Searls, 1994).

The general accuracy evaluation is based on the average of specificity and sensitivity:

$$AC = (Sn + Sp)/2$$

and correlation coefficient:

$$CC = (TP \times TN - FN \times FP)/[(TP + FN)$$
$$\times (TN + FP) \times (TP + FP) \times (TN + FN)]^{1/2}$$

One can see that the CC depends on the relation between the sites and non-functional sequences samples and can be important when measuring global prediction accuracy. However, the CC has an undesirable property: it is not defined when $TP + FN$, $TN + FP$, $TP + FP$ or $TN + FN$ is zero. To solve this problem, the average conditional probability (ACP) was suggested as an appropriate measure of global prediction accuracy (Burset and Guigo, 1996):

$$ACP = [TP/(TP + FN) + TP/(TP + FP) + TN/(TN + FP) + TN/(TN + FN)]/4$$

ACP ranges from 0 to 1. It is also possible to transform the ACP into AC (approximation coefficient):

$$AC = (ACP - 0.5) \times 2$$

In this way the AC range varies from -1 to 1 and can be compared with the CC.

All these measures have different statistical behaviour and can be used to compare various approaches.

4   FUNCTIONAL SITES IN NUCLEOTIDE SEQUENCES

Functional site analysis and recognition is based on the search for general context features common to all sites involved in a specific function. Usually, functional sites are recognized by specific factors. Invariant structures are very rarely found in sets of functional sites (for example, the AUG initiation site); a large variation is usually typical for functional site structure.

4.1   Binding Sites of Transcription Factors

Protein coding genes are transcribed by RNA polymerase II. Transcription is initiated in the promoter region by a complex of different factors. Computational analysis of transcriptional regulatory elements is an important field of human genome analysis. In recent years there has been a tremendous increase in experimental work devoted to understanding different aspects of transcription regulation. Different information about transcriptional sites and transcriptional factors is collected by several databases. The elementary sequence signals are typically short (in the range of 5 to 30 base pairs) and highly variable, reflecting the biochemistry of the decoding mechanism.

One of the early databases of transcription factors and signals (Faisst and Meyer, 1992) contained information only about transcription factors and the consensus sequences of their binding sites in vertebrate promoters. The consensus sequence is derived from a set of site sequences aligned with respect to functionally important position (e.g. transcription initiation site, exon–intron boundary). For each position a most conservative nucleotide is chosen. The simplest consensus generally consists of the most common nucleotides for each position. An example of consensus construction for ABF1 binding sites is shown in Figure 10.1. Other approaches to construct the consensus were analysed by Day and McMorris (1992).

The Transcriptional Factor Database (TFD) (Ghosh, 1993) and the compilation of DNA binding sites for protein transcription factors from vertebrates (Boulinkas, 1994) are generic collections of information about transcriptional binding sites. A more specific database devoted to MEF2 and myogenin sites has been developed by Fickett (1996b).

The most complete transcription factor database is TRANSFAC (Wingender *et al.*, 1996). This database consists of two parts: SITES and FACTORS. SITES is a collection of the description of the location of transcriptional sites within genes, functional properties of the sites, the type of the regulatory region (e.g. promoter, enhancer), and short comments on function. An example of the description is shown in Figure 10.2. FACTORS collects all the information on

```
GTCGTCTCACACG
ATCTTTGTTAACG
ATCGTTAATGACG
GTCACTGTACACG
GTCACGATATACG
ATCCCCATTAACG
ATCTCTCGCAACG
ATCATTATGCACG
ATCATTGAAAACG
GTCGTCTCACACG
-------------
ATCNTTNNNNACG - Consensus
```

Figure 10.1 *Example of consensus construction for ABF1 binding sites.*

```
ID   HS$A4_01
AC   R00010
DT   20.06.1990 (created); EWI.
DT   28.01.1991 (updated); THH (TFDB ENTRY).
TY   DNA
DE   A4 amyloid protein
SE   GGGCGCRGG.
SF   -200
ST   -100
OS   human, Homo sapiens.
OC   eukaryota;animalia; metazoa; chordata; vertebrata;
     tetrapoda; mammalia;
OC   eutheria; primates.
SO   0100; HeLa.
ME   gel retardation
RN   [1]
RA   Salbaum J. M., Weidemann A., Lemaire H.G.,
     Maters C. L., Beyreuther K.;
RT   The promoter of Alzheimer's disease amyloid A4
     precursor gene;
RL   EMBO J. 7:2807-2813 (1988).
DR   EMBL; X12751; HSPADP(3554:3562).
```

Figure 10.2 *Example of the HS$A4_01 entry from TRANSFAC database (Wingender et al., 1996).*

the transcription factors. This database can be found at the URL (http://trans-fac.gbf-braunschweig.de).

Major RNA polymerase II promoter elements are the TATA-box, the cap signal, the CCAAT box, and the GC box. The TATA box is found in the majority of protein coding genes. It is usually located about 25 to 30 nucleotides upstream from the transcription initiation site. It was suggested that this site is important for the accurate positioning of RNA polymerase II relative to the initiation site. All these signals are usually small and present a high variability. Their consensus can be described by using the IUPAC-IUB standard code or by the weight matrix. The IUPAC-IUB codes use a single character to describe the ambiguous nucleotides present in the consensus:

A, T(U), G, C,
R = (A or G),
Y = (T or C),
S = (G or C),
W = (A or T),
K = (G or T),
M = (A or C),
B = (T or G or C),
V = (A or G or C),
H = (A or T or C),
D = (A or T or G),
N = (A or T or G or C).

The weight matrices are derived from a set of aligned functional site sequences of length L. The weight matrix $F = |f(i, j)|$, (i = A, T, G, C; j = 1, L) is the nucleotide frequency matrix, and $f(i, j)$ is the absolute frequency of occurrence of the ith type of the nucleotide at the jth position. The optimized weight matrix of the functional site can be constructed in the following way:

$$W = |w(i, j)|, \qquad w(i, j) = \ln(f(i, j)/e(i, j) + s/100) + c(i)$$

where $e(i, j)$ is the expected nucleotide frequency of the ith type at the jth position, $c(i)$ is the column-specific free constant and s is the smoothing percentage. The weight matrix is then used to discriminate and recognize the functional sites above the threshold matching score value. More complicated weight matrices can be derived by various linear transformations (Berg and von Hippel, 1987; Frech *et al.*, 1993). The weight matrix and derived consensus sequence for the TATA box (Bucher, 1990) is shown in Figure 10.3. Consensus sequences for the cap signal, the CCAAT-box, and the GC-box are shown in Figure 10.4.

A different method used to predict the signal sites in the 5′ gene regions is the Hamming–Clustering network (Milanesi *et al.*, 1996). This approach employs a technique deriving from the synthesis of digital networks in order to generate prototypes, or rules, which can be directly analysed or used for the

Posi- tions	-3	-2	-1	0	+1	+2	+3	+4	+5	+6	+7	+8	+9	+10	+11
(a)															
A	16	4	90	1	91	69	92	57	40	14	21	21	21	17	20
C	37	12	0	2	0	0	1	1	11	35	38	33	30	28	26
G	39	5	1	1	1	0	5	11	40	39	33	33	33	36	36
T	8	79	9	96	8	31	2	31	9	12	8	13	16	19	18
(b)	G	T	A	T	A	A	A	A	G	G	C	G	G	G	G
	C		T		T	T		T	A	C	G	C	C	C	C

Figure 10.3 *Description of the initiation codon. (a) The percentage of each nucleotide for the eukaryotic TATA box positions (recalculated from numbers). (b) Consensus sequence (Bucher, 1990).*

Functional site	Sequence	Preferred location
CCAAT-box	Y Y Y R R C C A W W S R	-212 .. -57
GC-box	W R K R G G Y R K R K Y Y K	-164 .. +1
cap-site	K C W K Y Y Y Y	+1 .. +5

Figure 10.4 *Consensus sequences for the cap signal, the CCAAT box and the GC box (Bucher, 1990).*

construction of a final neural network. The Hamming–Clustering network technique has been successfully applied for TATA box analysis and prediction (Milanesi *et al.*, 1996) and the program can be used in the Internet at URL (http://www.itba.mi.cnr.it/webgene).

All DNA binding domains of transcription factors can be divided into several main classes: b/ZIP domains; bHLH domains; homeodomain proteins; EST domains; REL domains; zinc finger domains; HMG domains. Each class can be subdivided into subclasses. Transcription factors may differ significantly within each subclass. The main problem of transcriptional site analysis is the small volume of known sites for most types of transcription factor and the presence of closely related sequences (Frech *et al.*, 1993). Several programs for the recognition of transcriptional sites are integrated with databases of transcriptional sites and databases of derived consensus sequences and weight matrices of binding sites. Among them are: SIGNAL SCAN (Prestridge, 1991), MATRIX SEARCH (Chen *et al.*, 1995), ConsInspector (Frech *et al.*, 1993) and MatInspector (Quandt *et al.*, 1995). ConsInspector and MatInspector can be found at the Web sites (http://transfac.gbf-braunschweig.de; http://www.gsf.de). The use of these tools with different filters can significantly improve prediction accuracy (Quandt *et al.*, 1996). Analysis of transcriptional sites can help to determine gene function by defining which regulatory systems control its transcription (Bucher *et al.*, 1996).

4.2   Translation Initiation Sites

The AUG codon is the universal initiation site for eukaryotic translation (although very rare exceptions have been found). Analysis has shown that translation in about 95% of mRNA starts at the first AUG. This observation is in agreement with the scanning mechanism of translation initiation (Kozak, 1989), although the 'ribosome skipping model' (in which the ribosome may skip over certain regions of 5' UTR) cannot be excluded.

The analysis of initiation sites revealed a relatively weak sequence context. The consensus sequence GCCRCC/AUGG was suggested. The frequency matrix of the vertebrate initiation site is shown in Figure 10.5. The distribution of the

```
--------------------------------------------
                  5' UTR            CDS
Posi-   -6  -5  -4  -3  -2  -1  +1  +2  +3
tions
--------------------------------------------
(a)     G   C   C   A   C   C   A   T   G
                    G
(b)
--------------------------------------------
A       18  19  24  68  23  15 100   0   0
C       21  40  58   2  55  53   0   0   0
G       47  23  12  30  16  23   0   0 100
T       13  18   6   0   7   9   0 100   0
--------------------------------------------
```

Figure 10.5 *Description of the initiation codon. (a) Consensus sequence. (b) The percentage of each nucleotide for the vertebrate initiation site positions (Guigo et al., 1992).*

distance from mRNA start to the initiation AUG was used in the prediction of translation initiation sites (Guigo *et al.*, 1992). This region is relatively short (usually less than 200 nucleotides) (Hawkins, 1988).

4.3　Splice Site Analysis and Recognition

The splicing of introns is part of a multistep process of RNA maturation which takes place in the nucleus to generate mature mRNA molecules for transport to the cytoplasm. This process involves several factors such as snRNPs (small nuclear ribonucleoprotein particles) and hnRNPs (heterogeneous nuclear ribonucleoprotein particles). This complex assembly is called the spliceosome. It has been found that introns usually begin with GU (donor splice site) and end with AG dinucleotides (acceptor splice site). A short branch point signal in the intron sequence containing the adenosine residue is involved in the lariat structure formation. The branch point signal is located in introns upstream from acceptor splice sites. Several U snRNAs are involved in recognizing signal sequences: U1 snRNA recognizes the donor splice site, U2 snRNA presumably recognizes the branch site, and U5 snRNA interacts with the acceptor site (Maniatis and Reed, 1987). Numerous approaches have been suggested for the analysis and prediction of splice sites (for recent reviews see Gelfand, 1995; Rogozin and Milanesi, 1997).

4.3.1　The Donor Splice Site

A nine-nucleotide consensus (A,C)AG/GT(A,G)AGT (Figure 10.6a) was derived for the donor splice sites (Mount, 1982). The frequency matrix of human splice sites is shown in Figure 10.6b (Senapathy *et al.*, 1990). The results of recognition of the human donor splice sites with the use of the weight matrix approach are given in Table 10.1. In many cases, the real splicing sites have high homology scores (these splice sites are highly ranked, being at or near the top of the list). However, some real sites are ranked very low. This means that, by choosing high cut-off values of the matching score, it is possible to lose real sites. When lowering the cut-off value, the number of false splice sites (pseudosites) classified as potential splice sites will increase greatly.

The prediction accuracy of the weight matrix approach employed in the GeneID program (Guigo *et al.*, 1992) was calculated by using a test set of sequences (Rogozin and Milanesi, 1997). Results of prediction are shown in Table 10.2. One can see that accuracy of prediction by using the weight matrix is low. Many false splicing sites were found, although about 2–3% of donor splice sites were lost. This suggests that such an imperfect prediction can be due to the intrinsic limit of the weight matrix method. For example, the weight matrix

```
------------------------------------
         exon              intron
Posi-  -3 -2 -1      +1 +2 +3 +4 +5 +6
tions
------------------------------------
(a)    C  A  G  /   G  T  A  A  G  T
       A                G
------------------------------------

(b)
------------------------------------
   A   28 59  8  /   0   0 54 74  5 16
   C   40 14  5  /   0   0  2  8  6 18
   G   17 13 81  /100   0 42 11 85 21
   T   14 14  6  /   0100  2  8  4 45
------------------------------------
(c)
------------------------------------
          A  G  /  G  T  A
                /  G  T  A  A  G  T
          R  G  /  G  T  G  A  G
          A  G  /  G  T  N  N  G  T
------------------------------------
(d)
------------------------------------
          A  G  /  G  T  R
             G  /  G  T  N  A  G
                /  G  T  R  A  G
                /  G  T  R  N  G  T
             G  /  G  T  A
------------------------------------
```

Figure 10.6 *Description of the donor splice sites. (a) Consensus sequence from Mount (1982). (b) The percentage of each nucleotide for the human donor splice sites positions (Senapathy et al., 1990). (c) Four sequences from Iida and Sasaki (1983). (d) Five consensus sequences from Rogozin and Milanesi (1997).*

does not take into account the presence of a considerable number of correlated positions that had previously been revealed in donor site sets.

Iida and Sasaki (1983) have constructed a set of four sequence patterns for the donor splice sites (Figure 10.6c). They found that most donor sites contain one of these sequences and that such patterns rarely occur in exons. This idea was further developed by incorporating the method of functional site classification (Kudo *et al.*, 1992). Based on the classification approach, the splice site

Table 10.1   Splice site ranks for HUMMH sequence (Senapathy *et al.*, 1990)

Intron	Donor site			Acceptor site		
	Rank	Score	Sequence	Rank	Score	Sequence
1	12	75.1	GC/GTGAGT	7	91.4	CTCCTCGCTCCCAG/G
2	2	88.4	CG/GTGAGT	> 100	61.7	CGGGGGCGGGCCAG/G
3	> 100	60.8	GG/GTACCA	8	91.0	CTCTTTCCCGTCAG/A
4	3	87.6	GG/GTAAGG	11	87.6	CCCCCTTTTCCCAG/A
5	1	95.4	AG/GTAAGG	4	93.8	TTTTCTTCCCACAG/A
6	4	87.6	AA/GTAAGT	3	95.8	TTATTCTACTCCAG/G

prediction program SpliceView uses a set of five consensus sequences instead of a single consensus to describe the whole site set (Figure 10.6d). This program can be found at the ITBA CNR Web server at URL (http://www.itba.mi.cnr.it/webgene/). The results of donor site prediction using these methods are given in Table 10.2. One can see that the consensus approach (Rogozin and Milanesi, 1997) gives a more accurate description of donor sites than the weight matrix method.

Table 10.2   Errors of donor splice site prediction

Method (program)	Control set	E1	E2	Ac	CC	ACP	Reference
Weight	I	0.03	0.55	0.71	0.12	0.62	Guigo *et al.*
matrix	II	0.02	0.53	0.73	0.14	0.62	(1992)
(GeneId)	III	0.02	0.52	0.73	0.13	0.62	
Five consensus	I	0.05	0.15	0.90	0.30	0.73	Rogozin and
sequences	II	0.04	0.15	0.91	0.32	0.75	Milanesi (1997)
	III	0.04	0.15	0.91	0.30	0.73	
Consensus	I	0.05	0.18	0.89	0.27	0.72	Mount (1982)
MAG/GURAGU	II	0.06	0.18	0.88	0.29	0.72	
with three or	III	0.07	0.18	0.88	0.26	0.71	
fewer mismatches							
Discriminant	I	0.10	0.04	0.93	0.53	0.80	Solovyev *et al*
analysis of	II	0.11	0.03	0.93	0.60	0.82	(1994)
context	III	0.09	0.03	0.94	0.55	0.81	
factors (HSPL)							
Neural	I	0.29	0.02	0.85	0.51	0.77	Brunak *et al.*
network	II	0.28	0.03	0.85	0.50	0.77	(1991)
(NETGENE)	III	0.18	0.03	0.90	0.52	0.78	

More complicated techniques have been applied to splice site prediction. They generally take into account local characteristics of donor splice sites in combination with statistical properties of exons and introns. The neural network approaches had been applied for analysis of splice sites. Neural networks have several features that make them very useful for the analysis of biological sequences. They use both positive and negative information to train the neural network: sequences with the feature of interest (set of functional sites) and without the feature (set of pseudosites). They are able to detect complex correlations between positions in a functional site. Neural networks are able to determine which context features are important for recognition (for a review see Hirst and Sternberg, 1992). NETGENE is a neural network-based program for splice site prediction (Brunak *et al.*, 1991). This program is available via an e-mail server (netgene@virus.fki.dth.dk). The linear discriminant analysis used in the HSPL program (Solovyev *et al.*, 1994) (http://defrag.bcm.tmc.edu:9503/gene-finder/gf.html) takes into consideration local characteristics of the splice sites in combination with the statistical properties of exons and introns. The results of a comparison of these methods are shown in Table 10.2.

4.3.2 The Acceptor Splice Site

The short consensus CAG/G, which is preceded by a region of pyrimidine abundance (polypyrimidine tract), is a characteristic feature of acceptor sites. The consensus for the acceptor splice sites of primates is shown in Figure 10.7a,

					intron										exon
Posi-tions	-14	-13	-12	-11	-10	-9	-8	-7	-6	-5	-4	-3	-2	-1 /	+1
(a)	T	T	T	T	T	T	T	T	T	T	N	C	A	G /	G
		C	C	C	C	C	C	C	C	C					
(b)															
A	10	8	6	6	9	9	8	9	6	6	23	2	100	0	28
C	31	36	34	34	37	38	44	41	44	40	28	79	0	0	14
G	14	14	12	8	9	10	9	8	6	6	26	1	0	100	47
T	44	43	48	52	45	44	40	41	45	48	23	18	0	0	11

Figure 10.7 *Description of the acceptor splice sites. (a) Consensus sequence from (Senapathy et al., 1990). (b) The percentage of each nucleotide for the position of the human acceptor splice sites (Senapathy et al., 1990).*

Table 10.3   Errors of acceptor splice site prediction

Method (program)	Control set	E1	E2	Ac	CC	ACP	Reference
Weight	I	0.04	0.20	0.88	0.22	0.71	Guigo *et al.*
matrix	II	0.05	0.20	0.87	0.23	0.70	(1992)
(Geneld)	III	0.06	0.19	0.88	0.22	0.71	
Discriminant	I	0.14	0.02	0.92	0.56	0.80	Solovyev *et al.*
analysis of	II	0.25	0.02	0.87	0.56	0.79	(1994)
context	III	0.27	0.03	0.85	0.42	0.74	
factors (HSPL)							
Neural	I	0.20	0.07	0.86	0.31	0.72	Brunak *et al.*
network	II	0.13	0.06	0.91	0.42	0.76	(1991)
(NETGENE)	III	0.16	0.07	0.89	0.31	0.72	

and the frequency matrix is shown in Figure 10.7b (Senapathy *et al.*, 1990). The methodology for acceptor sites and for donor splicing site prediction is similar (see also Table 10.1). In Table 10.3 the accuracy of acceptor site prediction by using different programs is shown. One can see that prediction accuracy using the weight matrix for the acceptor is much higher than that with donor splice sites, and the HSPL and NETGENE programs decrease in prediction accuracy.

Although donor and acceptor splice sites are recognized by the same splicing machinery, the properties of these sites are very different.

4.3.3   The Branch Point Signal

The branch point signal is located within the range of 10–50 bases upstream from the acceptor splice site (the lariat region). The location of the functional branch sites is demonstrated experimentally in only a small number of cases. A five-nucleotide consensus sequence CT(G,A)A(C,T) is shown in Figure 10.8a and the nucleotide frequency matrix for the branch point signal derived from plant, rat, human, chicken and *Drosophila* sequences is shown in Figure 10.8b. The analysis of the consensus distribution along the gene sequences shows that the potential branch point sites are randomly distributed in exons and introns. Thus, the problem of an accurate branch point signal prediction is difficult to solve.

4.3.4   Prediction of Splice Sites in cDNA

Prediction of potential splice sites or intron-less subsequences in a cDNA is

```
--------------------
           intron
Posi-  -3 -2 -1  0 +1
tions
--------------------
(a)      C  T  G  A  C
                  A     T
--------------------
(b)
--------------------
A       1  0 39 99 11
C      76  8 15  1 45
G       2  0 42  0  6
T      21 91  4  0 38
--------------------
```

Figure 10.8   *Description of the branch point signal. (a) Consensus sequence from (Senapathy* et al., *1990). (b) The percentages of each nucleotide for human, rat, chicken, plant,* Drosophila *branch point signal positions (Senapathy* et al., *1990). 0 is the position of the 'branched nucleotide'.*

important for some experimental techniques. Former splice sites in cDNA consists of exonic parts of donor and acceptor splice sites. As a result, the consensus of former splice sites in a cDNA should be MAG/G.

The INTRON program has been developed for intron-less subsequence prediction (Globek *et al.*, 1991). For every input sequence INTRON performs three steps: it (1) finds all GG/CC pairs (possible former exon–intron splice sites); (2) finds all stop codons in all possible reading frames; and (3) eliminates the pairs that probably do not represent former splice sites based on an analysis of distance between these former sites and the closest stop codons.

More recently, another program called RNASPL has been developed (Solovyev *et al.*, 1994). This program is part of a complex of programs for promoter, 3'-processing, splice sites, coding exons and gene structure identification in genomic DNA. It is available at http://defrag.bcm.tmc.edu:9503/gene-finder/gf.html or by sending your file containing a sequence (a sequence name is in the first string) to service@bchs.uh.edu with the subject line 'rnaspl'.

4.3.5   Problems of Splice Site Prediction

Although good results were obtained for splice site prediction, the problem of correct splice site recognition remains. Splicing enhancer sequences have been found in exons adjacent to splice sites. mRNA secondary structure may play a key role in splice site recognition. It should be taken into account that splicing signals are evaluated by spliceosomes in relation to other splice sites, as has been shown in mutational experiments. The presence of the donor splice sites in the introns and exons is supported by mutationally occurring cryptic splice sites and cases of alternative splicing. Therefore, one cannot assume that splice sites are completely absent in the pseudosite sets.

A number of non-canonical splice sites has been revealed. The GC dinucleotide is found sometimes instead of GT for donor splice sites, although GC splice sites match well the standard MAG/GTRAGT consensus in other positions. Highly conserved N/ATATCCTT (for donor site) and TCCAC/W (for acceptor site) consensus sequences have been found in some genes. These non-canonical introns can be recognized by a specific class of factors (Hall and Padgett, 1994).

The problems of splice site prediction reflect the general problems in functional sites analysis, although splice sites are one of the best experimentally investigated signals and large datasets are available.

4.4   Polyadenylation Signal

Polyadenylation (cleavage of pre-mRNA 3' end and synthesis of poly(A) tract) is a very important early step of pre-mRNA processing. The most well known signal involved in this process is AATAAA, located 15–20 nucleotides upstream from the poly(A) site (site of cleavage) (Proudfoot, 1991). Real AATAAA signals can differ from AATAAA consensus sequence, although about 90% of mRNAs have a perfect copy of this sequence. The most frequent natural variant, ATTAAA, is nearly as active as the canonical sequence. Analysis of neighbouring bases showed that some other bases can be important for AATAAA recognition (Milanesi *et al.*, 1996). An additional signal with consensus YGTGTTYY (diffusive GT-rich sequence) was revealed in region from 20 to 30 nucleotides downstream of the poly(A) site (site of cleavage) (McLauchlan *et al.*, 1985). The weight matrix-based prediction is employed in the GRAIL system (Uberbacher *et al.*, 1996). Another program for polyadenylation site recognition (Kondrakhin *et al.*, 1994) is based on the matrix of three nucleotide frequencies. The Hamming–Clustering network technique has been also applied for poly(A) site prediction (Milanesi *et al.*, 1996) and the program is available at the WebGene server (http://www.itba.mi.cnr.it/webgene).

5   FUNCTIONAL REGIONS IN NUCLEOTIDE SEQUENCES

Functional sites in nucleotide sequences usually cannot be represented as islands in the sea of an irrelevant random background (Trifonov, 1996). Real nucleotide sequences are much more complex: functional sites must be in the right nucleotide context and can function in combination with other functional units forming functional regions.

5.1   Promoter Regions

It is well established that several factors can take part in the expression regulation of a eukaryotic gene. Each factor usually interacts with the binding site in DNA and with other factors. Several alternative pathways of regulation, and the synergism of multiple transcriptional factors and the great complexity of their structure, create serious problems when designing methods for the computational analysis of the eukaryotic polymerase II promoter regions.

Protein–protein interactions play an important role in transcription regulation. Such interactions depend on the composite elements containing binding sites for several factors. Information about composite regulatory elements can be found in the COMPEL database (Kel *et al.*, 1995a). Information regarding eukaryotic promoters is accumulated in the databases: EPD (Bucher, 1988), TRRD (Kel *et al.*, 1995b) and TRANSFAC (Wingender *et al.*, 1996). EPD is an annotated collection of experimentally characterized transcription initiation sites given as pointers to positions in EMBL Data Library sequences. The EPD database is available from the anonymous EMBL-EBI ftp site (ftp.ebi.ac.uk). The TRRD can be found on the IC&G Web server (http://www.bionet.nsc.ru). Both databases integrate with transcription factor and binding site databases TRANSFAC (http://transfac.gbf-braunschweig.de).

All these databases, in connection with the EBML and the SWISSPROT databases, are part of unintegrated database WEBGENDB. These databases can be interrogated at the ITBA CNR webserver (http://www.itba.mi.cnr.it/webgendb). The result will be represented graphically by an applet written in JAVA, as is shown in Figure 10.9 (Milanesi and D'Angelo, 1996).

GenomeInspector (http://www.gsf.de/biodv/genomeinspector.html) (Quandt *et al.*, 1996) is a software package for further interactive analysis of results obtained with the ConsInspector (Frech *et al.*, 1993) and MatInspector (Quandt *et al.*, 1995) programs. The analysis of distance correlations between large sets of potential functional sites can help in the identification of basic patterns of functional units. Several analysis modules allow visualization of pattern distributions and basic statistics, restriction of datasets by user-defined criteria, distance correlation analysis, r-scan statistics, extraction of compiled results and/or primary sequence data (Quandt *et al.*, 1996).

Several programs have been developed for the prediction of promoter regions. Most are based on the well known fact that binding sites of transcription factors are usually concentrated in the promoter region of eukaryotic genes, although they can also be found in other gene regions.

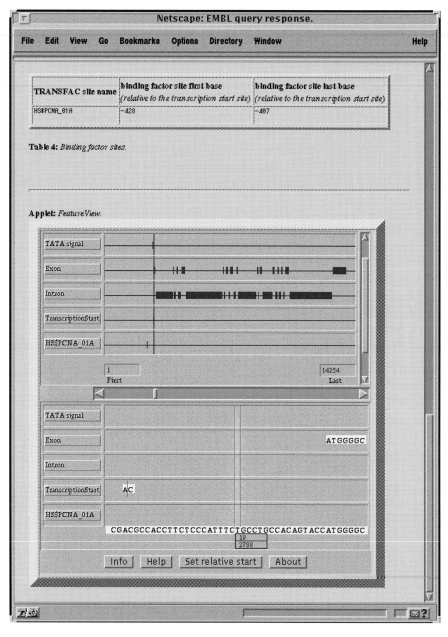

Figure 10.9   *Graphical representation of the EMBL, TRANFAC and EPD entry.*

```
HUMNTRIII
Promoter region predicted on reverse strand in 1382 to 1132
Promoter Score: 53.06 (Promoter Cutoff = 53.000000)
TATA found at 1148, Est.TSS = 1115
Significant Signals:
  Name              Strand   Location   Weight
  AP-2                +        1341      1.091000
  T-Ag                -        1333      1.086000
  Sp1                 -        1332      3.013000
  GCF                 -        1330      2.361000
  Sp1                 +        1327      3.061000
  UCE.2               +        1325      1.216000
  CTF                 -        1308      1.704000
```

Figure 10.10 *Text output of the PROMOTER SCAN program for human HUMNTRIII sequence.*

A PROMOTER SCAN program (Prestridge, 1995) has been developed to recognize promoter regions. Primate promoter sequences and non-promoter sequences have been used to compare the transcription site densities. The density of each of these binding sites is used to derive a ratio of density for each site between promoter and non-promoter sequences. The combination of the individual density ratios has been used to build a scoring profile. The example of the PROMOTER SCAN output is shown in Figure 10.10.

A FunSiteP program (Kondrakhin *et al.*, 1995) has been developed for revealing the eukaryotic promoter region. This method is based on the analysis of binding site distributions to construct the weight matrix of the specific binding sites location. The program is able to produce the profile of potential transcription regulatory regions.

Another approach for promoter recognition is based on the analysis of hexamer frequencies derived from promoter regions, coding regions and non-coding regions (Hutchinson, 1996). Different approaches are used in GRAIL, GeneId and other gene identification systems. In general, it is difficult to predict the promoter region owing to the high heterogeneity and complexity of these functional regions.

5.2 Protein Coding Regions

The evolution of coding sequences is strongly constrained by the encoded protein product. Statistical characteristics of protein coding regions are usually called coding potential or coding measures. Most coding potentials are based on analysis of codon usage (for review see Fickett and Tung, 1992; Gelfand, 1995). Different coding potentials have been compared by Fickett and Tung (1992). It has been concluded that a simple and obvious measure – counting oligonucleotides – is more effective than any of the more sophisticated methods. Claverie *et al.* (1990) used the differences between the frequency of occurrence of all hexamers between coding sequences and introns. The most successful coding measure is based on hexamers in phase with the coding frame (dicodons) (Fickett and Tung, 1992):

```
ATGCCATTGTTGCGATCC
ATGCCA
    CCATTG
        TTGTTG
            · · · · · · · · ·
```

For each possible dicodon ($i1 \ldots i6$), the frequency of occurrence in coding regions p($i1 \ldots i6$) and introns q($i1 \ldots i6$) has been calculated. The simplest discriminant score for coding region prediction is calculated by averaging d($i1 \ldots i6$) = p($i1 \ldots i6$)/(p($i1 \ldots i6$) + ($i1 \ldots i6$)) in a specific window length (Claverie *et al.*, 1990). The general accuracy of coding regions prediction by using dicodon statistics varied from 81% to 85% depending on window length.

For vertebrates the mean length of internal exons is 137 nucleotides (Hawkins, 1988).

5.3   Introns

The introns represent the non-coding region. There are several context features in intron sequences. The most important one is the branch point signal. The other is the high frequency of poly-G tracts near the 5′ end in many introns (Nussinov, 1988; Engelbrecht *et al.*, 1992; Solovyev *et al.*, 1994). It is possible to note the absence of AG in the 10 nucleotides preceding the AG at the intron–exon junction; and the GA and GG nucleotides in the region upstream of intron–exon boundary. These characteristics can be used for acceptor splice site recognition (Senapathy *et al.*, 1990).

One of the potential intron functions is to maintain the complex chromatin structure of protein coding genes. For example, the chromatin in cDNA (without introns) has been shown to be significantly different from that in the genomic DNA (Liu *et al.*, 1995).

Vertebrate genes have introns with a wide range of length, although short ones (80–99 nucleotides) predominate. About 20% of introns are longer than 1600 nucleotides and the mean length of introns is 1127 nucleotides (Hawkins, 1988).

5.4   3' Untranslated Regions

Although the AATAAA-GT-rich signal represents the minimum context requirement for poly(A), other elements upstream of the polyadenylation site may also
influence its efficiency (for review see Wahle, 1995). This property was used for
the discriminant analysis of 3' UTR context features to predict the 3' processing regions (Salamov and Solovyev, 1997). The accuracy has been estimated for
the set of 131 poly(A) regions and 1466 non-poly(A) regions of human genes
having the AATAAA sequence. For 86% accuracy poly(A) region prediction
(Sn = 0.86), the algorithm has 8% false predictions (CC = 0.62). It is available
at http://defrag.bcm.tmc.edu:9503/gene-finder/gf.html or by sending your file
containing a sequence (a sequence name is in the first string) to
service@bchs.uh.edu with the subject line 'polyah'. The Hamming–Clustering
network technique has also been applied successfully for poly(A) prediction
(Milanesi *et al.*, 1996). The program can be found on the Internet at the URL
http://www.itba.mi.cnr.it/webgene.

Most vertebrate 3' UTRs have a length of 100–700 nucleotides, although long
3' UTRs (length greater than 1500 nucleotides) can be also found (Hawkins, 1988).

5.5   Repeated Regions

A large part of the human genome consists of repeated DNA. Repeated DNA
can be roughly divided into tandemly and dispersed repeated DNA elements.

The classical human satellites I–IV (comprising 2–5% of the haploid genome)
are made up of tandemly repeated DNA (short reiterated sequences of fewer
than 100 nucleotides containing divergent repeats of the pentamer GGAAT).
The other family of tandemly repeated DNA is the alphoid family of satellite
DNA (comprising 4–6% of the haploid genome). Tandemly repeated DNA is
present at the centromeres of human chromosomes as well as in other locations.
It participates in different aspects of mitotic chromosome movements and is
associated with different proteins (for review see Pluta *et al.*, 1990).

Dispersed repeated elements are presented by different retroposons, which can
be divided into viral and non-viral superfamilies based on common structural
features. The length of non-viral retroposons varies significantly (from 30 to
6000–8000 nucleotides). Almost all short interspersed elements (SINEs) are
derived from known RNA polymerase III transcripts such as 7SL RNA and
tRNA. Long interspersed elements (LINEs) are transcribed by RNA polymerase
II and contain long open reading frames (ORFs).

The (CCG)n and (CAG)n repeats expansion (dynamic mutations) can cause
several diseases in humans. Very large triplet expansions lead to fragile sites on
chromosomes (for review see Willems, 1994).

```
HUM
;Human test sequence
;SQ    Sequence 290 BP;  68 A;  82 C;  96 G;  44 T;  0 other;
TEST
ggccgggcgcggtggctcacgcctgtaatcccagcactttgggaggccgaggcgg
gcggatcacctgaggtcaggagttcgagaccagcctggccaacatggtgaaaccc
cgtctctactaaaaatacaaaaattagccgggcgtggtggcgcgcgcctgtaatc
ccagctactcgggaggctgaggcaggagaatcgcttgaacccgggaggcggaggt
tgcagtgagccgagatcgcgccactgcactccagcctgggcgacagagcgagact
ccgtctcaaaaaaaa1
```

Figure 10.11 *Example of the CENSOR input.*

Repeated elements can have a number of functions: they can be recombination hotspots; they can be incorporated in the translated parts of host genes; they can participate in gene regulation; or they can serve as a source of polyadenylation signals (for review see Makalowski, 1995). The presence of repeated elements can create serious problems for sequence analysis especially in the case of homology searches in nucleotide sequence databases. Repeated elements must be treated carefully by using special programs and databases.

The CENSOR e-mail server allows users to align query sequence(s) against a reference collection of human or rodent repeats (Jurka *et al.*, 1992, 1996). The homologous regions are then 'censored'. Censoring means replacing the homologous regions with X in the query sequences. The server automatically classifies Alu repeats and adds the classification to the report. The e-mail server is (censor@charon.lpi.org). The query sequence(s) must be in Stanford/IG format. An example of an input file is shown in Fig. 10.11. The server currently censors human and rodent sequences. The CENSOR program and the database of repetitive elements (repbase) can be found on the anonymous ftp site (ftp://ncbi.nlm.nih.gov/repository/repbase/).

When predicting SINE elements based on homology with SINE consensus sequences, it must be taken into account that many SINEs originated from tRNA sequences (for review see Weiner *et al.*, 1986). Because of this, the tRNA genes may be censored during repeated elements analysis and it is not possible to reveal the tRNA genes in further analysis.

6 PROTEIN CODING GENE STRUCTURE PREDICTION

6.1 Problems of Gene Structure Prediction

The different methodologies for revealing splice sites and coding regions alone are not able to predict the gene structure. The coding region prediction methods miss most of the short exons and cannot reliably define the exon–intron boundaries, whereas the splice site prediction reveals some of real splice sites together with a great number of false sites. The main approach for gene structure prediction combines information about local functional sites (splice sites, initiation codon) together with global features of coding regions and introns. Although there is no direct biological relation between statistical properties of protein coding regions (at the mRNA translation level) and splicing (at the pre-mRNA processing level), such combinations can strongly improve gene structure prediction. Additional information can be obtained from prediction of promoters (at the DNA transcription level) and the 3' UTRs (at the pre-mRNA processing level).

To predict the protein coding exons and genes, a number of programs have been developed (for review see Milanesi *et al.*, 1994; Fickett, 1995, 1996a; Gelfand, 1995; Snyder and Stormo, 1995a; Burset and Guigo, 1996; Fickett and Guigo, 1996; Uberbacher *et al.*, 1996). All these programs are based on the combination of the potential functional signals with global statistical properties of protein coding regions (Fields and Soderlund, 1990; Gelfand, 1990; Uberbacher and Mural, 1991; Guigo *et al.*, 1992; Hutchinson and Hayden, 1992; Borodovsky and McIninch, 1993; Gelfand and Roytberg, 1993; Milanesi *et al.*, 1993; Snyder and Stormo, 1993, 1995b; Dong and Searls, 1994; Solovyev *et al.*, 1994; Thomas and Skolnick, 1994; Kulp *et al.*, 1996). The performance of some of these programs on well defined test sets has shown that the average portion of missing protein coding exons ranges between 0.22 and 0.36, whereas the average portion of false exons ranges between 0.13 and 0.27, having the CC between 0.60 and 0.80 (Burset and Guigo, 1996). These results show that all these programs are far from being able to reveal the complete structure of protein coding genes. Moreover, the gene prediction program's accuracy was supposed to be lower for large genomic sequences having complex structure in comparison with relatively short sequences, which are usually used as control sets (Burset and Guigo, 1996). A comprehensive computer analysis of long chromosome regions will require the development of new methods able to predict complex gene structures.

Another approach to gene prediction methodology is based on the homology detection throughout the databases of nucleotide or amino acid sequences (Guigo *et al.*, 1992; Borodovsky *et al.*, 1994; States and Gish, 1994; Brown *et al.*, 1995; Snyder and Stormo, 1995b; Gelfand *et al.*, 1996a; Rogozin *et al.*, 1996; Uberbacher *et al.*, 1996). By using the information available on homologous protein sequences, it is possible to improve the accuracy of gene structure prediction significantly.

Real sequences can contain artifactual nucleotide substitutions, insertions and deletions, as many sequencing strategies are error prone. This fact should be taken into account in sequence analysis. The investigation of sequences with introduced frameshift mutations has shown that the results of gene prediction programs suffer substantially (Burset and Guigo, 1996). Several algorithms have been developed for the analysis of frameshift errors (Fichant and Quentin, 1995;

Xu *et al.*, 1995) based on statistical properties of coding sequences. The accuracy of error correction is not high, but an improvement in prediction accuracy for cases with a high frameshift rate has been shown. Revealing errors can be improved significantly by using information on homologous proteins. Error correction techniques should be used carefully, as there are many pseudogenes in eukaryotic genomes, and straightforward computer correction should be confirmed by experimental procedures.

The current generation of gene recognition tools is very inefficient at recognizing the beginning and end of genes (Bucher *et al.*, 1996). This may not be a problem for the analysis of only one gene. With large-scale sequencing, the inability to separate several genes has become a serious problem. As a result, tools for prediction of promoter and poly(A) regions has become one of the most important parts of current systems.

Further analysis of predicted protein functions and structure can help to confirm the gene prediction results, although such analysis is difficult for proteins without significant homologies to known proteins already present in the databases.

Currently available collections of expressed sequence tags (ESTs) are large and thus useful for gene mapping (Boguski *et al.*, 1994). Homology searches against the EST Division of GenBank (dbEST) can be used for this purpose. The exact matches in dbEST correspond to one or more EST matches in dbEST which have at least 97% identity with the query (http://www.ncbi.nlm.nih.gov/Bassett/dbEST/). By using a lower identity score, the presence of repeated elements in the 5' and 3' UTR must be taken into account.

CpG island prediction can be important for gene identification, as CpG islands are often considered to be gene marks and are frequently found at the 5' ends of genes. Other cases of genome inhomogeneity should be taken into account, especially A + T content (Gelfand, 1995; Guigo and Fickett, 1995; Uberbacher *et al.*, 1996). Genome inhomogeneity can be responsible for a very low prediction accuracy for some genes: for example, GRAIL revealed three true protein coding sequences and 37 non-real protein coding sequences in the human retinoblastoma gene (27 CDS in 180 388 nucleotides) (Thomas and Skolnick, 1994).

Most of the gene structure prediction programs have been developed for human and vertebrate genomes. The methodology of many programs described here can also be applied to different organisms. A short description of systems available on the Internet is shown in Table 10.4. To show output formats from some of these programs we used the human HUMNTRIII (AC L12691) sequence. This sequence includes two CDS in positions 2627–2801 and 3382–3491.

6.2   GRAIL

GRAIL (Gene Recognition and Analysis Internet Link) is an interface to a computer system that provides the prediction of different functional regions

Table 10.4   Gene identification systems and programs

System	Availability	Result	Error revealing	Using homology	5' and 3' ends revealing	Analysis of potential proteins
GRAIL	E-mail server XGRAIL client–server system	Exons Genes	+	+	+	+
FGENEH	E-mail server Web server	Exons Genes	–	–	+	++
WebGene	Web server	Exons Genes	+	+	+	–
GeneId	E-mail server	Genes	–	+	+	–
GeneParser	Package	Genes	Low sensitivity	+	–	–
GeneMark	E-mail server	Exons	–	–	–	–
PROCRUSTES	Web server	Genes	+	+	–	–
GenLang	E-mail server	Genes	–	–	–	–
Genie	Web server	Genes	–	+	–	–
SORFIND	Package	Genes	–	–	–	–
GREAT	Package	Genes	–	–	–	–
Xpound	Package	Genes	–	–	–	–
MZEF	Package	Exons	–	–	–	–

(protein coding sequences, promoters, poly(A) signals, CpG islands, etc.) in DNA sequences of human, mouse, *Arabidopsis*, *Drosophila* and *E. coli* genomes. The protein coding recognition module (CRM) is based on a multisensor neural network (Uberbacher and Mural, 1991; Uberbacher *et al.*, 1996).

GRAIL is available on the e-mail server (grail@ornl.gov) or on the Web server (http://avalon.epm.ornl.gov/Grail-1.3). The X-window client–server system (XGRAIL) can be used for an interactive mode of sequence analysis. The User's Guide and programs are available from the anonymous ftp (arthur.epm.ornl.gov).

The prediction methods in the GRAIL system are based on the Coding Recognition Module (CRM) consisting of a group of seven sensor algorithms. The program scans the sequence by a fixed window, and the seven statistical tests (frame bias matrix, Fickett statistic, fractal dimension, coding hexamers, in-frame hexamers (dicodons), k-tuple commonality, repetitive hexamers) are calculated for each window. Each value constitutes the input node to the neural net, which is able to calculate the potential coding value.

An example of the output is shown in Figure 10.12. The Final Exon Prediction output gives the limits of the extent of the coding exons, the most likely strand,

```
HUMNTRIII, len = 3710

Exon predication on forward strand:

start/acceptor donor/stop    rf     score           orf
-------------------------------------------------------------
    2101      --    2144      3       70       2061  -  2330
    2615      --    2801      2       92       2561  -  2923
    3382      --    3491      3       66       3312  -  3491

Exon predication on reverse strand:

start/donor acceptor/stop    rf     score           orf
-------------------------------------------------------------
     997      --    1124      2       40        884  -  1156


Final Exon Predication:

start/donor acceptor/stop strand  rf    quality         orf
-------------------------------------------------------------
    2101      --    2144      f      3   excellent  2061  -  2330
    2615      --    2801      f      2   excellent  2561  -  2923
    3382      --    3491      f      3   excellent  3312  -  3491
```

Figure 10.12 *Text output of GRAIL system for human HUMNTRIII sequence.*

the preferred reading frame for the exon and a quality assessment (excellent, good or marginal).

The GRAIL has been tested by the authors as well as by other researchers (Burset and Guigo, 1996). The program was able to reveal 90% of coding exons of a length of more than 100 nucleotides.

6.3   FGENEH

FGENEH (Solovyev *et al.*, 1994) predicts internal exons, 5' and 3' exons by linear discriminant functions analysis applied to the combination of various contextual features of these exons. The optimal combination of these exons is calculated by the dynamic programming technique to construct the gene models. The accuracy has been estimated by the authors using a set of 193 complete human gene sequences. The accuracy of precise exon recognition was 81% with the Sp (specificity) of 79%, while the Sn (sensitivity) at the level of individual nucleotides was 90% with the Sp = 93%. For exon prediction in partially sequenced genes

```
HUMNTRIII.
length of sequence -   3710
number of predicted exons -  4
positions of predicted exons:
  1476 -    1628 w=   1.51
  2061 -    2144 w=   6.50
  2615 -    2801 w=  13.51
  3382 -    3491 w=   5.41
Length of Coding region - 534bp  Amino acid sequence - 177aa
MSSKKTITGLWNILLRETYLGPWSVFQSGCLIFAIELLEFLMYSDICPFCHPDKIIVLST
LIPTGDYSPHNLKNLFMRMVTPAMRTLAILAAILLVALQAQAEPLQARADEVAAAPEQIA
ADIPEVVVSLAWDESLAPKHPGSRKNMDCYCRIPACIAGERRYGTCIYQGRLWAFCC*
```

Figure 10.13   *Output of FGENEH program for human HUMNTRIII sequence.*

it is better to use FEXH (5′, internal and 3′ exon prediction) and HEXON (internal exon prediction) programs. All these programs are also available for the *Drosophila, Caenorhabditis elegans* and *Arabidopsis* genes. This set of programs is available at http://defrag.bcm.tmc.edu:9503/gene-finder/gf.html. An example of the output file is shown in Figure 10.13.

6.4   WebGene

WebGene is a computer system for in silico gene structure analysis for: promoters, poly(A) signals, splice sites prediction and gene structure identification. Programs are available for human, mouse, *Drosophila*, *C. elegans*, *Arabidopsis* and *Aspergillus* genes. The prediction methods (GenView, ORFGene, HC-TATA, HC-PolyA, SpliceView, etc.) can be used in combination with different sequence databases during the gene discovery process. The WebGene system is available on the Internet at the ITBA CNR Web server (http://www.itba.mi.cnr.it/webgene).

The GenView program (Milanesi *et al.*, 1993) contains the following procedures: (1) identification of the potential splice sites; (2) construction of a potential coding fragment (PCF); (3) estimation of the potential of the revealed PCF; (4) construction of potential genes having the maximal coding potential (mode GENE); and (5) obtaining the best PCFs selection (mode EXON). For the splice site prediction the classification analysis (Rogozin and Milanesi, 1997) has been used. A second step performs a verification of the potential splice sites using an alternative technique based on the weight matrix (Senapathy *et al.*, 1990). From the set of potential splice sites, stop codons and potential AUG codons, all possible PCFs are constructed. To estimate the protein coding potential of revealed PCF, a dicodon statistics is used (Claverie *et al.*, 1990). It is possible to use the program to predict a full gene structure (GENE option) or to find the best potential exons (EXON option).

With the GENE option, the set of revealed PCFs is used to construct the potential genes with a maximal coding potential. The search for a set of potential genes with maximal coding potential is based on the dynamic programming technique. For the GenView (v. 2.0), the comparison between the control dataset and the results of the prediction shows that 72% of real exons and 4% false exons were predicted (CC 0.75). The GENE option can be used to reveal the whole, or part, of a gene.

With the EXON option only the highly significant PCFs are selected. Two levels of significance for the best PCFs are used: excellent and good PCF. For the 'excellent exon' mode 30% of true exons was predicted and no false exons were found, and for the 'good exon' mode 50% of true exons and 2% of false exons were found (CC 0.70). The EXON option can be especially useful for long genomic sequences with an unknown number of potential genes, as there is a very small overprediction.

The ORFGene program (Rogozin *et al.*, 1996) implements gene structure prediction by using information on homologous proteins. The ORFGene system can significantly improve the results of the techniques for gene prediction based on simple characteristics of coding sequences, especially if highly homologous proteins are present in the amino acid sequence database. The program involves the following procedures: (1) starting from the potential peptides predicted by the GenView (2.0) program, it is possible to reveal the key amino acid sequence through the exhaustive searching for homologies between the potential protein sequence and the protein database; (2) construction of the complete set (W1) of PCFs and the complete set (R1) of potential peptides (PPs); (3) selection of potential peptides from R1 based on the homology with the chosen amino acid sequence; (4) selection of a set of 'best' PCFs (W2) for the gene structure reconstruction; (5) reconstruction of potential genes on the base of the compatibility graph Q by using a dynamic programming technique; and (6) detecting, in the set of potential genes, the one with the best homology with the key amino acid sequence.

The accuracy of this program depends on the homology level between the key sequence and the protein sequence. In the case of high homology with the key protein it is possible to achieve 95% and higher accuracy (AC).

To correct the potential sequencing errors in the input sequence, a procedure for frameshift correction has also been included in the GenView (2.0) and the ORFGene programs.

The result of the analysis is presented in EMBL-like format and by using a special graphical interface to visualize the gene structure predicted on the Internet. The applet FeatureView (Milanesi and D'Angelo, 1996) was implemented in JAVA and can be used for the interactive result evaluation in the WebGene system. An example of the text output file is shown in Figure 10.14.

```
ID   HUMNTRIII
XX
OS   Homo sapiens
XX
CC   Search in direct strand
XX
CC   Results of gene prediction by ORFGENE computer system
XX
FH   Key          Location/Qualifiers
FH
FT   CDS          2625..2810
FT                /note=potential exon predicted by ORFGENE system
FT                /note=based on homology with P11479
FT                /note=internal exon
FT   CDS          3382..3488
FT                /note=potential exon predicted by ORFGENE system
FT                /note=based on homology with P11479
FT                /note=last exon
XX
CC   Translated peptide
CC   MRTLAILAAILLVALQAQAEPLQARADEVAAAPEQIAADIPEVVVSLAWDESLAPKHPGE
CC   RGSRKNMDCYCRIPACIAGERRYGTCIYQGRLWAFCC
XX
```

Figure 10.14 *Text output of the ORFGene program for human HUMNTRIII sequence.*

6.5   GeneId

GeneId (Guigo *et al.*, 1992) is a hierarchical rule base computer system for gene prediction in DNA sequences.

The GeneId e-mail server is geneid@darwin.bu.edu. GeneId was originally designed for predicting the exon–intron structure in full-length pre-mRNA. In the case that the sequence does not contain first or last exons, then GeneId will try to predict the first and last exons. The analysed sequence must be no less than 100 and no more than 20 000 nucleotides.

At the first step of analysis GeneID is able to reveal the sets of all potential sites: initiation codons, stop codons, acceptor and donor splice sites. A set of potential first, internal and terminal coding exons is then generated. Predicted exons are filtered by a special set of rules. Series of variables are determined for each predicted exon. In accordance with the values of these variables, the potential exon may be rejected. The goal is to reduce the size of the set of potential exons by discarding as many false exons as possible, without discarding the true predicted ones. Some 24 different variables are computed for each predicted exon. These variables are:

1. Variables 1–4. Fraction of nucleotides A, T, C and G in the exon sequence.
2. Variables 5–8. Codon position correlations: χ^2 statistic of correlation between the third position of a codon and the first position of the next one, and between the middle position of two consecutive codons, has been found to be characteristic of coding sequences.
3. Variables 9–16. Numeric derivatives of variables 1–8 at the beginning of the exon.
4. Variables 17–24. Numeric derivatives of variables 1–8 at the end of the exon.

A discriminant function partially based on neural network was derived for reducing the number of false predicted exons.

Finally, predicted genes are obtained as linear arrangements of exon classes. Each class consists of equivalent exons. Two exons are equivalent if they occur exactly in the same potential gene (i.e. they are completely interchangeable).

The example of the output is shown in Figure 10.15. The lists of potential start codons, stop codons, donor splice sites, acceptor splice sites, first exons, internal exons, last exons and reconstructed genes are returned.

GeneId has been tested on a set of 28 sequences. The sensitivity (Sn) in detecting true exons was 0.69, where the portion of the predicted coding nucleotides that are actually coding was 0.84 (Sen2).

6.6   GeneParser

GeneParser is a computer program for the identification of coding regions in genomic DNA sequences (Snyder and Stormo, 1993, 1995a,b). This program is based on several statistics: codon usage, local compositional complexity, length distribution of exons and introns, k-tuple frequencies, and weight matrices for splice signals. Dynamic programming was integrated with training a feedforward neural network to maximize the number of correct exons predictions.

An example of the output is shown in Figure 10.16. Detailed information regarding different statistics is presented.

GeneParser has been tested by using GeneId and GRAIL test sets of sequences. The sensitivity (Sn) was 0.73 and the specificity (Sp) was 0.71.

6.7   GeneMark

The GeneMark (Borodovsky and McIninch, 1993) has been set up for the

```
-----------------------------------------------------------------------
REPORT for genomic DNA sequence HUMNTRIII3710BPDNAP, 3710 bases.
Thu Sep  5 00:50:37 EDT 1996

0. Potential PolII PROMOTERS
Promoter prediction:
  Region      Score  Likelihood
    0 -  300  0.551  Marginal prediction
.......................................
 2900 - 3300  0.640  Marginal prediction

1. potential START CODONS
    Position  Score  Sequence
       90     0.67   TTGAATGG
.......................
     3644     0.66   TGATATGT

2. potential STOP CODONS
    Position  Codon
       11     TAA
............
     3700     TGA

3. potential DONOR SITES
    Position  Score  Sequence
        9     0.76   CCCTGTAAGCC
..........................
     3674     0.61   ACACGTACCAA

4. potential ACCEPTOR SITES
    Position  Score  Sequence
       26     0.75   GTAAGCCCTGTTACAGG
.................................
     3541     0.73   GCTTTGAGAGCTACAGG
-----------------------------------------------------------------------
5a. potential FIRST EXONS (from startcodon to donor site)
     Score  Frame  Start  Stop        Blastx matches    Blast score
     -----Highly confident predictions, more than 50% true exons: ---
     0.77    0            2142  2144
..........................................................
6a. potential INTERNAL EXONS
..........................................................
7a. potential LAST EXONS (from acceptor site to stopcodon)
..........................................................
-----------------------------------------------------------------------
5b. potential FIRST EXON CLASSES
    (Only exons upstream of 3'-most predicted last exon)
..........................................................
-----------------------------------------------------------------------
8. potential GENES
1132 gene models were analyzed and ranked according to likelihood
The 20 most likely genes are

Ranking scores   List of exons (*) constituting gene (first, internal(s), last)
  0.806  1.229    2142  2144     2615  2801     3108  3140     3382  3491
.......................................................................
  0.738  0.749     155   162      388   426      832   892     2615  2801
                  3108  3140     3382  3491
```

Figure 10.15 *Output of Geneld system for human HUMNTRIII sequence. Only part of the file is given.*

```
GeneParser:                    *********** T-matrix Score *************
Position      Class  RF  Donor  Accep   IF-6    LCC    Len    6-tup   L-mat

   10   2100 intron  0  0.875  0.825  -2.114  0.778  0.030  0.056  +0.408
 2101   2144 exon    2  0.875  0.947   0.504  0.747  0.410  0.501  -0.202
 2145   2614 intron  2  0.962  0.947   0.067  0.772  0.420  0.389  +0.275
 2615   2801 exon    1  0.962  0.947   0.574  0.810  0.620  0.428  -0.110
 2802   3107 intron  0  0.860  0.947   0.053  0.770  0.610  0.424  +0.255
 3108   3140 exon    1  0.860  0.947   0.501  0.806  0.410  0.494  -0.194
 3141   3381 intron  1  0.920  0.947   0.285  0.755  0.940  0.466  +0.266
```

```
        Position   beginning and ending nucleotides of sequence inclusive
        Class      sequence class, intron or exon
        RF         Exon: reading frame relative to beginning of sequence
                   Intron: intron length mod 3
        Donor      donor site
        Accep      acceptor site
        IF-6       in-frame 6-tuple log-likelihood
        LCC        local compositional complexity
        Len        Exon and intron length distribution
        6-tup      Exon and intron 6-tuple log-likelihood
        L-mat      L-matrix score for subsequence
```

Figure 10.16   *Output of GeneParser system for human HUMNTRIII sequence.*

identification of the protein coding regions in DNA sequences in different prokaryotic and eukaryotic species.

The GeneMark e-mail server (genemark@ford.gatech.edu) accepts a formated message containing a DNA sequence in text format with word 'genemark' in the Subject line. You may specify a species name and parameters that control the analysis procedure.

GeneMark is based on inhomogeneous and homogeneous Markov chains models for protein coding sequences and introns. GeneMark can be especially useful for finding prokaryotic genes, analysing cDNA sequences and discriminating coding regions from non-coding ones like 3' or 5' UTRs. This program is able to identify rather long exons (more than 100 nucleotides) in eukaryotic genomic DNA. This information can be used to design reverse transcriptase–polymerase chain reaction (RT-PCR) or PCR primers. The GeneMark program has been used for gene identification in *E. coli*, *Bacillus subtilis* and *Haemophilus influenzae* genome sequencing projects. An example of the output is shown in Figure 10.17.

6.8   PROCRUSTES

PROCRUSTES (Gelfand *et al.*, 1996a) is based on the spliced alignment algorithm which explores all possible exon assemblies and finds the gene structure

```
                       GENEMARK PREDICTIONS
Query:humntriii.seq
Sequence file: /var/tmp/caaa001wI
Sequence length: 3710
GC Content:  45.63%
Window length: 96
Window step: 12
Threshold value: 0.500
---
Matrix: H. sapiens, 0.00 < GC < 0.46 - Order 4
Matrix author: JDM
Matrix order: 4

List of Regions of interest
(regions from stop tp stop codon w/ coding function >0.500000)

   LEnd      REnd     Strand       Frame
 --------  --------  -----------  -----

      235       771  direct        fr 1
     1605      1784  complement    fr 2
     2558      2923  direct        fr 2

List of Open reading frames
(regions from start to stop codon w/ coding function >0.500000)

Left       Right     DNA          Coding Avg   ORF   Start
end        end       Strand       Frame  Prob  Prob  Prob
--------   --------  -----------  -----  ----  ----  ----

     1605      1730  complement    fr 2   0.60  0.95  0.20

List of Protein-Coding Exons
(regions between acceptor and donor site w/ coding function >0.500000)

  Left      Right
  End        End      Strand       Frame   Prob
 -------   -------   -----------  -----   ------

      291       421  direct        fr 1    0.8148
      291       407                        0.8584

     1654      1729  complement    fr 2    0.7277
     1662      1729                        0.8045

     2104      2250  direct        fr 3    0.4085
     2141      2212                        0.4691

     2632      2801  direct        fr 2    0.6377
     2675      2758                        0.8721

     2965      3038  complement    fr 2    0.5026
     2973      3024                        0.5246

     3421      3488  direct        fr 1    0.5438
```

Figure 10.17   *Text output of GeneMark system for human HUMNTRIII sequence.*

with the best fit to a related protein. PROCRUSTES considers all possible chains of candidate exons and finds a chain with the maximum global similarity to the target protein. The selection of the target protein in the PROCRUSTES depends on user choice. Error-tolerant gene recognition is also provided. This Web server (http://www-hto.usc.edu/software/procrustes/) can also be used for the recognition of highly specific exons and PCR primers selection (CASSANDRA).

6.9 GenLang

GenLang (Dong and Searls, 1994) is based on generative grammars and expresses the information in DNA sequences in a declarative hierarchical manner. It is implemented in a Prolog-based Definite Clause Grammar system, which is used to parse the DNA sequence for the analysis of genetic information. Sequence can be submitted to GenLang e-mail server by sending e-mail to genlang@cbil.humgen.upenn.edu. The format of the mail body should look like this:

```
NAME mytestseq
SPECIES {drosophila,vertebrate,dicot}
GENE {protein,tRNA}
SEQUENCE
gatctatattat...
. . . . . . . . . . . . . .
```

6.10 Genie

The Genie program (Kulp *et al.*, 1996) is based on a statistical model of genes in DNA. A Generalized Hidden Markov Model (GHMM) provides the framework for describing the rules of a DNA sequence parsing. Probabilities are assigned to transitions between states in the GHMM and to the generation of each nucleotide base, given a particular state. Machine learning techniques were applied to optimize these probabilities using a training set. The best parse of a sequence is produced from the model using a dynamic programming algorithm to identify the path through the model with maximum probability. The GHMM is flexible and modular, so new sensors and additional states can easily be inserted. In addition, it provides simple solutions for integrating reading frame constraints, errors of sequencing and homology searching. The exon sensor is a codon frequency model conditioned on windowed nucleotide frequency and the preceding codon. Two neural networks are used, as in Brunak *et al.* (1991), for splice site prediction. For a test set of 304 genes in human DNA, Genie identified up to 85% of protein coding bases correctly with a specificity (Sp) of 80%. Genie is available on the Web server (http://www-hgc.lbl.gov/projects/genie.html).

6.11   SORFIND

SORFIND (Hutchinson and Hayden, 1992) is the computing program for the prediction of internal exons in human genome sequences. Potential coding exons are stratified according to the reliability of their prediction from confidence level 1 to level 5.

After reading the sequence from the file, the program scans it from left to right, stopping at each AG dinucleotide. The site is then scored according to the method suggested by Berg and von Hippel (1987), and is accepted if its score is over the threshold for acceptor sites. The position of the first downstream stop codon in each of the three reading frames is then noted, and all GT dinucleotides that are at least 60 nucleotides downstream and within this window are revealed. Each potential donor site is scored, and the site is accepted if its score is over the donor threshold value. If a given sequence segment passes through the selection procedure, it must consist of at least one ORF with 60 nucleotides or more, bracketed by acceptor and donor splice sites. Such a structure is defined as a spliceable open reading frame (SORF). The program then calculates three separate variables, based upon codon usage, for each SORF. The best candidates among the SORFs are selected.

The output (Figure 10.18) is a list of potential internal exons, ordered by the confidence level, including start and stop positions, length, splice and codon usage scores, 5′ and 3′ phase and amino acid translations. The 5′ phase is defined as the number of nucleotides required from a previous exon to put the current SORF into its correct reading frame. Similarly, the 3′ phase is defined as the number of nucleotides from the current SORF carried over to the next exon. If the input is an annotated file in GenBank format, the output also includes a line for each SORF identifying its relationship to the exons described in the feature table.

The control set (14 genes) has been analysed. Specificity (the percentage of SORFs that correspond to true exons) varied from 91% at confidence level 1 to 16% at level 5.

6.12   GREAT

The GREAT (Genome Recognition and Exon Assembly Tool) program is based on vector dynamic programming (Gelfand and Roytberg, 1993; Gelfand *et al.*, 1996b). Scoring function takes into account codon usage and positional nucleotide frequencies of splicing signals. Sensitivity is 0.88 and specificity is 0.79. The program is available from the authors.

```
SORFIND   Version 1.5.1

Start Run: Thu Jan 03 08:57:03 1980

SORF:    349..574     Length: 226
Confidence: 3
Phase  5': 0   3': 2
Acceptor:   ctcctatgcctcccaga Score:   4.9
Donor   :          caggtgata Score:   4.03
Codon Usage          5'difference: 0.354
                     3'difference: 0.126
                     Average    : 0.028

Amino Acid Translation:
KAPLKLLLNASLKAHKAERLCNTSSKVNAQTPTSFLGWPSGKATPTLMANASDQFL
AQMILDNCLSLNCSLAKQT

SORF:    1561..1628    Length: 68
Confidence: 3
Phase  5': 2   3': 0
Acceptor:   tctgtttttcaatcagg Score:   5.63
Donor   :          catgtaggt Score:   4.13
Codon Usage          5'difference: 0.349
                     3'difference: 0.174
                     Average    : -0.051

Amino Acid Translation:
CLIFAIELLEFLMYSDICPFCH

SORF:    2615..2801    Length: 187
Confidence: 2
Phase  5': 0   3': 2
Acceptor:   gctctccctcctccagg Score:   2.36
Donor   :          caggtgaga Score:   1.14
Codon Usage          5'difference: 0.238
                     3'difference: 0.114
                     Average    : 0.066

Amino Acid Translation:
VTPAMRTLAILAAILLVALQAQAEPLQARADEVAAAPEQIAADIPEVVVSLAWDES
LAPKHP

Input File: 4.sta      3710 bp
GC Content: 0.46
```

Figure 10.18 *Output of SORFIND system for human HUMNTRIII sequence.*

6.13   Xpound

The Xpound program is based on the probabilistic Markov model. The results are presented in terms of the probability for each nucleotide in a sequence to be coding. This program runs locally on a Sun workstation and is freely available from A. Thomas (Thomas and Skolnick, 1994).

6.14   MZEF

MZEF (Zhang, 1997) is an internal coding exon prediction program. MZEF is compiled for SunOS 5.5 and mzef.osf is compiled for DEC Alpha OSF/1 3.2c. The program is available at the ftp site phage.cshl.org in pub/science/mzef.

7   ANALYSIS OF POTENTIAL PROTEINS CODED BY PREDICTED GENES

The most popular approach for the analysis of potential proteins is homology searches in protein databases (PIR, SWISSPROT, etc.). Potential protein predicted by using gene identification tools can be analysed by standard programs for homology searches (FASTA, BLAST) (Pearson and Lipman, 1988; Altschul *et al.*, 1990). Instead of this procedure the BLASTX program could be used (Gish and States, 1993). However, the latter is quite time consuming as it is based on the testing of all peptide fragments potentially coded by the sequence of interest. The BLASTC program (States and Gish, 1994) combines sequence similarity and codon bias information.

Another approach for potential protein analysis can be found on the SBASE e-mail server (sbase@icgeb.trieste.it). A database search is performed against the SBASE library of protein domains, and the search results, provided with a graphic display and the annotations on request, are returned in a mail message. The HSPCRUNCH (Sonnhammer and Durbin, 1994) program can be used for a graphic representation of the individual domain homologies.

If no homology is found for potential protein, then analysis of potential proteins becomes extremely complicated as significant errors in prediction are typical in gene identification tools (Burset and Guigo, 1996). However, it can be suggested that, in some cases, analysis of potential proteins can give important information about potential functions. Different approaches for such analysis can be applied.

Analysis of functional motifs in amino acid sequences is an interesting approach. Information about functional sites and patterns in proteins is accumulated in the PROSITE database (Bairoch, 1991). This database consists of the two main files, PROSITE.DAT and PROSITE.DOC, which contain different information about sites in computer-readable and free formats respectively. The PSITE program used for revealing patterns in protein sequences can be found on the WWW (http://dot.imgen.bcm.tmc.edu: 9331/pssprediction/pssp.html). The PSITE is based on statistical estimation of the expected number of a PROSITE pattern in a given sequence. It uses the PROSITE database of functional motifs. If a pattern that has an expected number significantly less than 1 is found, it can be supposed that the analysed sequence possesses the pattern function.

Secondary structure and three-dimensional structure prediction can also help in the analysis of potential function of a predicted protein. Expressed secondary structure elements can confirm reality of prediction. A large set of such programs can be found on several servers: Baylor College of Medicine (http://defrag.bcm.tmc.edu:9503/lpt.html), SOPMA server (http://www.ibcp.fr) and EMBL server (PredictProtein@embl-heidelberg.de).

8 RNA-CODING GENE STRUCTURE PREDICTION

Several approaches have been developed for the prediction of tRNA genes (for review see Gelfand, 1995). They are based on the recognition of highly conservative primary and secondary tRNA structures. tRNA genes are transcribed by RNA polymerase III. Internal promoters of RNA polymerase III have a highly conservative structure. A program for tRNA gene prediction can be found on the GenLang e-mail server (genlang@cbil.humgen.upenn.edu).

The prediction of other RNA genes is usually based on homology searches. Ribosomal RNA genes are transcribed by RNA polymerase I. Other RNA genes are transcribed by RNA polymerases II and III. The prediction of functional RNA genes is an open problem in computational biology, as a number of RNA pseudo-genes can be found in eukaryotic genomes. Homology searches can be performed by using these e-mail servers: BLAST program (blast@ncbi.nlm.nih.gov), FASTA program (fasta@ebi.ac.uk), PROSRCH program (dapmail@ed.ac.uk) and MPSRCH program (blitz@embl-heidelberg.de).

9 SEQUENCE ANALYSIS

In conclusion, we will suggest a possible 'scheme' to follow for sequence analysis. It is complicated to formalize the main steps of sequence analysis (functional mapping), because each sequence has unique features and thus needs specific investigation. However, we will try to describe some possible approaches for gene structure prediction, as this is the most common aspect of functional mapping.

Step 1. Revealing repeated elements. This is an important step because the presence of repeated elements can create problems in sequence analysis. Long repeated elements contain ORFs that can be recognized by gene identification tools as potential genes. It is complicated to interpret database search results if the output is saturated by a number of highly scored matches with repeated elements in nucleotide sequence databases.

Step 2. Homology searches in databases is an important point in functional mapping, because significant homology with some known functional region is the most obvious sequence landmark (and the only evidence in favour of reality of sequence functionality). In many computational tools, database searches are integrated as part of the system. A BLASTX search can be useful for revealing protein coding regions.

Step 3. If a highly homologous protein sequence (or EST) is revealed, then the gene structure will be reconstructed with high accuracy, although alternative splicing and the presence of other genes in the analysed sequence should be taken into account. Potential errors of sequencing can also be revealed with high accuracy at this step.

Step 4. If no homologous protein was revealed (steps 2 and 3), then the gene structure can be predicted by using coding statistics and potential functional motifs (splicing signals, initiation codons).

Step 5. Revealing potential transcriptional binding sites, promoters and poly(A) signals can help in understanding of the functional meaning of the analysed sequence. Analysis of CpG islands is important for genic regions recognition. Steps 4 and 5 are integrated in some systems.

Step 6. The analysis of homologies between potential peptides and proteins from databases. This is important for cases of weak, but significant, similarities. Such similarities can be lost during a BLASTX search, as all possible translations are used by this program (as a result, sensitivity will be lower). Further analysis of protein secondary structure and functional motifs can confirm the reality of revealed gene. Steps 3–6 can be repeated several times if more than one gene is present in a sequence.

There are many existing open problems described in this chapter. However, it can be concluded that nowadays a large variety of computational tools that can be very useful for gene structure analysis and prediction does exist. Each computational tool has its weak and strong points, and no unique 'excellent' program or system can really be found (for example, some programs can be very successful in gene structure prediction of pre-mRNA and some other programs for long uncharacterized DNA sequences). It is also better to use several programs for analysing the same sequence, as combining different approaches can improve prediction accuracy (Fickett and Tung, 1992; Burset and Guigo, 1996).

We hope that this chapter can help investigators of gene identification tools to select a suite of programs and servers for maximal 'coverage' to solve their problems.

ACKNOWLEDGEMENTS

The authors are grateful to D. D'Angelo for realization of the graphical interface of Webgene and to M. Bishop for helpful suggestions for revision. This work was supported by EC project BIO4-CT95-0226 and the CNR Italian-Russian 'Bioinformatics' Project.

REFERENCES

Altschul SF, Gish W, Miller W, Myers EW and Lipman DJ (1990) Basic local alignment search tool. *J. Mol. Biol.* **215**, 403–410.

Antequera F and Bird A (1993) Number of CpG islands and genes in human and genes. *Proc. Natl. Acad. Sci. U.S.A.* **90**, 11 995–11 999.

Bairoch A (1991) PROSITE: a dictionary of sites and patterns in proteins. *Nucleic Acids Res.* **19 (Supplement)**, 2241–2245.

Berg OG and von Hippel PH (1987) Selection of DNA binding sites by regulatory proteins. I. Statistical–mechanical theory and application to operators and promoters. *J. Mol. Biol.* **193**, 723–750.

Bernardi G (1989) The isochore organization of the human genome. *Annu. Rev. Genet.* **23**, 637–661.

Bernardi G and Bernardi G (1991) Compositional patterns of nuclear genes from cold-blooded vertebrates. *J. Mol. Evol.* **33**, 57–67.

Boguski MS, Tolstoshev CM and Bassett DE Jr (1994) Gene discovery in dbEST. *Science* **265**, 1993–1994.

Borodovsky M and McIninch J (1993) GenMark: parallel gene recognition for both DNA strands. *Comput. Chem.* **17**, 123–133.

Borodovsky M, Koonin EV and Rudd KE (1994) New genes in old sequences: strategy for finding genes in the bacterial genome. *Trends Biochem. Sci.* **19**, 309–313.

Boulinkas T (1992) Evolutionary consequences of nonrandom damage and repair of chromatin domains. *J. Mol. Evol.* **35**, 156–180.

Boulinkas T (1994) A compilation and classification of DNA binding sites for protein transcription factors from vertebrates. *Crit. Rev. Euk. Gene Expr.* **4**, 117–321.

Brown NP, Whittaker AJ and Newell WR (1995) Identification and analysis of multi-gene families by comparison of exon fingerprints. *J. Mol. Biol.* **249**, 342–359.

Brunak S, Engelbreacht J and Knudsen S (1991) Prediction of human mRNA donor and acceptor sites from the DNA sequences. *J. Mol. Biol.* **220**, 49–66.

Bucher P (1988) The Eukaryotic Promoter Database. *EMBL Nucleotide Sequence Data Library Release* **17**, Postfach 10.2209, D-6900 Heidelberg.

Bucher P (1990) Weight matrix description of four eukariotic RNA polymerase II promoter elements derived from 502 unrelated promoter sequences. *J. Mol. Biol.* **212**, 563–578.

Bucher P, Fickett JW and Hatzigeorgiou A (1996) Computational analysis of transcriptional regulatory elements: a field in flux. *Comput. Appl. Biosci.* **12**, 361–362.

Burset M and Guigo R (1996) Evaluation of gene structure prediction program. *Genomics* **34**, 353–367.

Chen QK, Hertz GZ and Stormo GD (1995) MATRIX SEARCH 1.0: a computer program that scans DNA sequences for transcriptional elements using a database of weight matrices. *Comput. Appl. Biosci.* **11**, 563–566.

Claverie JM, Sauvaget I and Bougueleret L (1990) k-Tuple frequency analysis: from intron/exon discrimination to T-cell epitope mapping. *Methods Enzymol.* **183**, 237–251.

Day WHE and McMorris FR (1992) Critical comparison of consensus methods for molecular sequences. *Nucleic Acids Res.* **20**, 1093–1099.

Dong S and Searls DB (1994) Gene structure prediction by linguistic methods. *Genomics* **23**, 540–551.

Engelbrecht J, Knudsen S and Brunak S (1992) G + C-rich tract in 5' end of human introns. *J. Mol. Biol.* **227**, 108–113.

Faisst S and Meyer S (1992) Compilation of vertebrate-encoded transcription factors. *Nucleic Acids Res.* **20**, 3–26.

Fichant GA and Quentin Y (1995) A frameshift error detection algorithm for DNA sequencing projects. *Nucleic Acids Res.* **23**, 2900–2908.

Fickett JW (1995) The gene identification problem: an overview for developers. *Comput. Chem.* **20**, 103–118.

Fickett JW (1996a) Finding genes by computer: the state of the art. *TIG* **12**, 316–320.

Fickett JW (1996b) Coordinate positioning of MEF2 and myogenin sites. *Gene* **172**, GC19–GC32.

Fickett JW and Guigo R (1996) Computational gene identification. In *Internet for the Molecular Biologist* (eds S Swindell, R Miller and G Myers), Horison Scientific Press, Oxford.

Fickett JW and Tung CS (1992) Assessment of protein coding measures. *Nucleic Acids Res.* **20**, 6441–6450.

Fields CA and Soderlund CA (1990) GM: a practical tool for automating DNA sequence analysis. *Comput. Appl. Biosci.* **6**, 263–270.

Frech K, Herrmann G and Werner T (1993) Computer-assisted prediction, classification, and delimitation of protein binding sites in nucleic acids. *Nucleic Acids Res.* **21**, 1655–1664.

Gelfand MS (1990) Computer prediction of the exon–intron structure of mammalian pre-mRNAs. *Nucleic Acids Res.* **18**, 5865–5869.

Gelfand MS (1995) Prediction of function in DNA sequence analysis. *J. Comput. Biol.* **1**, 87–115.

Gelfand MS and Roytberg MA (1993) Prediction of the exon–intron structure by a dynamic programming approach. *BioSystems* **30**, 173–182.

Gelfand MS, Mironov AA and Pevzner PA (1996a) Gene recognition via spliced sequence alignment. *Proc. Natl. Acad. Sci. U.S.A.* **93**, 9061–9066.

Gelfand MS, Podolsky LI, Astakhova TV and Roytberg MA (1996b) Recognition of genes in human DNA sequences. *J. Comput. Biol.* **3**, 223–234.

Ghosh D (1993) Status of the transcription factors database. *Nucleic Acids Res.* **21**, 3117–3118.

Gish W and States DJ (1993) Identification of protein coding regions by database similarity search. *Nature Genet.* **3**, 266–272.

Globek A, Gorski M and Polymeropoulos H (1991) *INTRON version 1.1 manual.* Laboratory of Biochemical Genetics, National Institute of Mental Health, Neuroscience Center at St Elizabeths, Washington, DC 20032.

Guigo R and Fickett JW (1995) Distinctive sequence features in protein coding, genic noncoding, and intergenic human DNA. *J. Mol. Biol.* **253**, 51–60.

Guigo R, Knudsen S, Drake N and Smith T (1992) Prediction of gene structure. *J. Mol. Biol.* **225**, 141–157.

Hall SL and Padgett RA (1994) Conserved sequences in a class of rare eukaryotic nuclear introns with nonconsensus splice sites. *J. Mol. Biol.* **239**, 357–365.

Hawkins JD (1988) A survey on intron and exon length. *Nucleic Acids Res.* **16**, 9893–9905.

Hirst JD and Sternberg MJE (1992). Prediction of structural and functional features of protein and nucleic acid sequences by artificial neural network. *Biochemistry* **31**, 7211–7218.

Hutchinson GB (1996) The prediction of vertebrate promoter regions using different hexamer frequency analysis. *Comput. Appl. Biosci.* **12**, 391–398.

Hutchinson GB and Hayden MR (1992) The prediction of exons through an analysis of spliceable open reading frames. *Nucleic Acids Res.* **20**, 3453–3462.

Iida Y and Sasaki F (1983) Recognition patterns for exon–intron junctions in higher organisms as revealed by a computer search. *J. Biochem.* **94**, 1731–1738.

Jurka J, Walichiewicz J and Milosavljevic AJ (1992) Prototypic sequences for human repetitive DNA. *J. Mol. Evol.* **35**, 286–291.

Jurka J, Klonowski P, Dagman V and Pelton P (1996) CENSOR – a program for identification and elimination of repetitive elements from DNA sequences. *Comput. Chem.* **20**, 119–122.

Karlin S, Blaisdell BE, Sapolsky R, Cardon L and Burge C (1993) Assessement of DNA inhomogeneities in yeast chromosome III. *Nucleic Acids Res.* **21**, 703–711.

Kel OV, Romaschenko AG, Kel AE, Wingender E and Kolchanov NA (1995a) A compilation of composite regulatory elements affecting gene transcription in vertebrates. *Nucleic Acids Res.* **23**, 4097–4103.

Kel OV, Romaschenko AG, Kel AE, Naumochkin AN and Kolchanov NA (1995b) Structure of data representation in TRRD – database of transcription regulatory regions on eukaryotic genomes. In *Proceedings of the 28th Annual Hawaii International Conference on System Sciences (HICSS), v.5 Biotechnology Computing,* IEE Computer Society Press, Los Alamitos, pp. 42–51.

Kondrakhin YV, Kel AE, Kolchanov NA, Romaschenko AG and Milanesi L (1995) Eukaryotic promoter recognition by binding sites for transcription factors. *Comput. Appl. Biosci.* **11**, 477–488.

Kozak M (1989) The scanning model for translation: an update. *J. Cell. Biol.* **108**, 229–241.

Kudo M, Kitamura-Abe S, Shimbo M and Iida Y (1992) Analysis of context of 5′-splice site sequences in mammalian pre-mRNA by subclass method. *Comput. Appl. Biosc.* **8**, 367–376.

Kulp D, Haussler D, Reese MG and Eeckman FH (1996) A generalized hidden Markov model for the recognition of human genes in DNA. In *Proceedings of the Fourth International Conference on Intelligent Systems for Molecular Biology*, AAAI Press, Menlo Park, pp. 134–142.

Li WH and Graur D (1991) *Fundamentals of Molecular Evolution*. Sinauer, Sunderland, MA.

Liu K, Sandgren EP, Palmiter RD and Stein A (1995) Rat growth hormone gene introns stimulate nucleosome alignment *in vitro* and in transgenic mice. *Proc. Natl. Acad. Sci. U.S.A.* **92**, 7724–7728.

Lopez RS, Larsen F and Prydz H (1994) Evaluation of the exons prediction of the GRAIL software. *Genomics* **24**, 133–136.

Makalowski W (1995) SINEs as a genomic scrap yard: an essay on genomic evolution. In *The Impact of Short Interspersed Elements (SINEs) on the Host Genome* (ed. RJ Maraia), R.G. Landes Company, Austin, TX, pp. 81–104.

Maniatis T and Reed R (1987) The role of small nuclear ribonucleoprotein particles in pre-mRNA splicing. *Nature* **325**, 673–678.

McLauchlan J, Gaffney D, Whitton JL and Clements JB (1985) The consensus sequence YGTGTTYY located downstream from the AATAAA signal is required for efficient formation of mRNA 3′ termini. *Nucleic Acids Res.* **13**, 1347–1369.

Milanesi L and D'Angelo D (1996) FeatureViewer: a graphical representation of the EMBL feature sequence entry. In *Proceedings of HUGO 1996 Conference*, Heidelberg, Germany.

Milanesi L, Kolchanov NA, Rogozin IB, Ischenko IV, Kel AE, Orlov YL, Ponomarenko MP and Vezzoni P (1993) GenViewer: a computing tool for protein-coding regions prediction in nucleotide sequences. In *Proceedings of the Second International Conference on Bioinformatics, Supercomputing and Complex Genome Analysis* (eds HA Lim, JW Fickett, CR Cantor and R Robbins), World Scientific, Singapore, pp. 573–588.

Milanesi L, Kolchanov NA, Rogozin IB, Kel AE and Titov I (1994) Sequence functional inference. In *Guide To Human Genome Computing* (ed. MJ Bishop), Academic Press, London, pp. 249–312.

Milanesi L, Muselli M and Arrigo P (1996) Hamming-clustering method for signals prediction in 5′ and 3′ regions of eukaryotic genes. *Comput. Appl. Biosci.* **13**, 399–404.

Mount SM (1982) A catalogue of splice junction sequences. *Nucleic Acids Res.* **10**, 459–472.

Nussinov R (1988) Conserved quartets near 5′ intron junctions in primate nuclear pre-mRNA. *J. Theor. Biol.* **133**, 73–84.

Pearson, WR and Lipman DJ (1988) Improved tools for biological sequence comparison. *Proc. Natl. Acad. Sci. U.S.A.* **85**, 2444–2449.

Pluta AF, Cooke CA and Earnshaw WC (1990) Structure of the human centromere at metaphase. *Trends Biochem.* **15**, 181–185.

Prestridge DS (1991) SIGNAL SCAN: a computer program that scans sequences for eukaryotic transcriptional elements. *Comput. Appl. Biosci.* **7**, 203–206.

Prestridge DS (1995) Predicting Pol II promoter sequences using transcription factor binding sites. *J. Mol. Biol.* **249**, 923–932.

Proundfoot NJ (1991) Poly(A) signals. *Cell* **64**, 671–674.

Quandt K, Frech K, Karas H, Wingender E and Werner T (1995) MatInd and MatInspector: new fast and versatile tools for detection of consensus matches in nucleotide sequence data. *Nucleic Acids Res.* **23**, 4878–4884.

Quandt K, Grote K and Werner T (1996) GenomeInspector: a new approach to detect correlation patterns of elements on genomic sequences. *Comput. Appl. Biosci.* **12**, 405–413.

Rogozin IB and Milanesi L (1997) Analysis of donor splice sites in different eukaryotic organisms. *J. Mol. Evol.* **45**, 50–59.

Rogozin IB, Milanesi L and Kolchanov NA (1996) Gene structure prediction using information on homologous protein sequence. *Comput. Appl. Biosci.* **12**, 161–170.

Salamov AA and Solovyev VV (1997) Recognition of 3'-processing sites of human mRNA precursors. *Comput. Appl. Biosci.* **13**, 23–28.

Senapathy P, Shapiro MB and Harris NL (1990) Splice junctions, branch point sites, and exons: sequence statistics, identification and Genome Project. *Methods Enzymol.* **183**, 252–278.

Solovyev VV, Salamov AA and Lawrence CB (1994) Predicting internal exons by oligonucleotide composition and discriminant analysis of spliceable open reading frames. *Nucleic Acids Res.* **22**, 5156–5163.

Sonnhammer ELL and Durbin R (1994) A workbench for large scale sequence homology analysis. *Comput. Appl. Biosci.* **1**, 301–307.

Snyder EE and Stormo GD (1993) Identification of coding regions in genomic DNA sequences: an application of dynamic programming and neural networks. *Nucleic Acids Res.* **21**, 607–613.

Snyder EE and Stormo GD (1995a) Identifying genes in genomic DNA sequences. In *DNA and Protein Sequence Analysis: A Practical Approach*, 2nd edn (eds MJ Bishop and CJ Rawlings), pp. 209–223. IRL Press, Oxford.

Snyder EE and Stormo GD (1995b) Identification of protein coding regions in genomic DNA. *J. Mol. Biol.* **248**, 1–18.

Staden R (1984) Computer methods to locate signals in nucleic acid sequences. *Nucleic Acids Res.* **12**, 505–519.

Staden R (1985) Computer methods to locate genes and signals in nucleic acids sequences. In *Genetic Engineering, Principle and Methods v7* (eds JK Setlow and A Hollaender), Plenum Press, New York, pp. 67–114.

States DJ and Gish W (1994) Combined use of sequence similarity and codon bias for coding region identification. *J. Comput. Biol.* **1**, 39–50.

Stormo GD (1988) Computer methods for analyzing sequence recognition of nucleic acids. *Ann. Rev. Biophys. Biophys. Chem.* **17**, 241–263.

Thomas A and Skolnick MK (1994) A probabilistic model for detecting coding regions in DNA sequences. *IMA J. Math. Appl. Med. Biol.* **11**, 149–160.

Trifonov EN (1996) Interfering contexts of regulatory sequence elements. *Comput. Appl. Biosci.* **12**, 423–430.

Uberbacher EC and Mural RJ (1991) Locating protein coding segments in human DNA sequences by a multiple sensor-neural network approach. *Proc. Natl. Acad. Sci. U.S.A.* **88**, 11 261–11 265.

Uberbacher EC, Xu Y and Mural RJ (1996) Discovering and understanding genes in human DNA sequence using GRAIL. *Methods Enzymol.* **266**, 259–281.

Wahle E (1995) 3'-End cleavage and polyadenylation of mRNA precursor. *Biochim. Biophys. Acta* **1261**, 183–194.

Weiner AM, Deininger PL and Efstratiadis A (1986) Nonviral retroposons: genes, pseudogenes, and transposable elements generated by the reverse flow of genetic information. *Ann. Rev. Biochem.* **155**, 631–661.

Willems PJ (1994) Dynamic mutations hit double figures. *Nature Genet.* **8**, 213–215.

Wingender E, Dietz P, Karas H and Knuppel R (1996) TRANSFAC: a database on transcription factors and their DNA binding sites. *Nucleic Acids Res.* **24**, 238–241.

Xu Y, Mural RJ and Uberbacher EC (1995) Correcting sequencing errors in DNA coding regions using a dynamic programming approach. *Comput. Appl. Biosci.* **11**, 117–124.

Zhang MQ (1997) Identification of protein coding regions in the human genome based on quadratic discriminant analysis. *Proc. Natl. Acad. Sci. U.S.A.* **94**, 565–568.

11 Gene Finding: Putting the Parts Together

Anders Krogh

Center for Biological Sequence Analysis, Technical University of Denmark, Building 206, 2800 Lyngby, Denmark

1 INTRODUCTION

Any isolated signal of a gene is hard to predict. Current methods for promoter prediction, for instance, will have either a very poor specificity or a very poor sensitivity, such that they will either predict a huge number of false positives (fake promoters) or a very small number of true promoters. The same is essentially true for splice site prediction: if looked at in isolation, splice sites are hard to recognize with good accuracy. This may seem like a contradiction, because there are programs that perform well on this task such as those of Brunak *et al.* (1991) and Solovyev *et al.* (1994). The reason for the success is that both these methods also use the statistics of the coding exon region next to the splice site. Apart from doing a very careful job of describing the regions right around the splice site, they can therefore also rule out splice sites that do not sit next to what appears to be a good coding region. In bird-watching, the surroundings often give the necessary clues in deciding which bird you are watching in the distance; whether it is seen in an open field or in a wood, for instance. Signal detection in genes is much like bird-watching: it is necessary to take the surroundings into account.

Therefore, to predict a splice site it is also necessary to predict coding exons, and vice versa (disregarding the splice sites of introns in untranslated regions). In a long DNA sequence, one probably would not expect to see a coding exon with two associated splice sites unless there were other exons with which it could combine. In this way predictions of the various parts of a gene should influence each other, and prediction of the entire gene structure will also improve on the predictions of the individual signals. Therefore, in the past few years, gene prediction has moved more and more towards prediction of whole gene structures, and these methods typically use modules for recognition of coding regions, splice

Guide to Human Genome Computing, 2nd edition   Copyright © 1998 Academic Press Limited
ISBN 0-12-102051-7

sites, translation initiation and termination sites, and some even use statistics of the 5′ and 3′ untranslated regions (UTRs), promoters, etc. This combination of predictions has indeed improved the accuracy of gene prediction considerably, and as more knowledge is gained about transcription and translation it is likely that the integration of other signals can improve it even further.

In this paper I will describe some of the methods for combining the predictions of several signals into a prediction of a complete gene structure. Because most of the gene finding methods are reviewed in Chapter 10, I will focus on the methods used for combining sensors.

2 DYNAMIC PROGRAMMING

A variety of methods has been proposed for combining predictions from various sensors into one predicted gene structure. Some of the early methods are GeneModeler (Fields and Soderlund, 1990) and GeneID (Guigo *et al.*, 1992), which both predict candidate exons and then combine them according to various rules. Many programs use some sort of dynamic programming for the combination (Gelfand and Roytberg, 1993; Snyder and Stormo, 1993, 1995; Stormo and Haussler, 1994; Xu *et al.*, 1994; Solovyev *et al.*, 1995; Gelfand *et al.*, 1996; Wu, 1996). Genefinder also uses dynamic programming (P. Green and R. Durbin, personal communication). A simple dynamic programming algorithm will be given first, which represents the essence of many of the algorithms, such as those in Snyder and Stormo (1993), Stormo and Haussler (1994) and Wu (1996). The emphasis will be on the overall principles rather than the rigorous details.

Assume the following functions, which all score a region between base number i and number j in the sequence at hand:

- ■ internal_exon(i,j) scores the region for being an internal coding exon.
- ■ initial_exon(i,j) scores the region for being the coding part of the first coding exon, which may include a score for the translation initiation signal a few bases upstream of the ATG start codon.
- ■ terminal_exon(i,j) scores the region for being the coding part of the last coding exon. A score for translation end is included.
- ■ single_exon(i,j) scores the region for being the coding part of an unspliced gene including scores for translation start and stop.
- ■ intron(i,j) scores the region for being an intron. This includes the scores for the splice sites, even if the splice site detectors use the first or last few bases of the neighbouring exons outside the interval from i to j.

■ `intergenic(i,j)` scores the region for being intergenic. Such a sensor is often omitted, which corresponds to setting the score to zero, so we include it without loss of generality. Note that, despite the name, this function is also used for scoring the 5′ and 3′ UTR of the gene. Explicit incorporation of UTR sensors will be addressed later.

Usually the exon score is calculated by adding up some sort of hexamer statistics along the sequence from i to j, as in, for example, GeneParser (Snyder and Stormo, 1993, 1995) and FGENEH (Solovyev *et al.*, 1995), but it can also be neural network based as in GRAIL (Xu *et al.*, 1994) and Genie (Kulp *et al.*, 1996, 1997). However, it is also possible to use database hits to score exons, which is used in GeneParser, PROCRUSTES (Gelfand *et al.*, 1996), Genie and GeneWise (Birney and Durbin, 1997), for example.

One widely used method for scoring coding regions is the inhomogeneous Markov chain of the nth order, which is used extensively in GeneMark (Borodovsky and McIninch, 1993). In the simplest case (first order), the 16 conditional probabilities $p_k(a \mid b)$ for nucleotide a, given the previous one (b) in the sequence, are needed for the three positions in the reading frame $k = 1, 2, 3$. These can be estimated from a set of known coding regions simply by counting the occurrence of the 16 dinucleotides in each frame and normalizing properly. To score a sequence $x_i, \ldots, x_j$ as being coding and starting in frame 1, one multiplies all these probabilities along the sequence, $p_1(x_i \mid x_{i-1}) \times p_2(x_{i+1} \mid x_i) \times \cdots$. By taking the logarithm, this score becomes additive. If we let $lp_k(a \mid b) = \log p_k(a \mid b)$, the score is:

$$lp_1(x_i \mid x_{i-1}) + lp_2(x_{i+1} \mid x_i) + lp_3(x_{i+2} \mid x_{i+1}) + lp_1(x_{i+3} \mid x_{i+2})$$
$$+ lp_2(x_{i+4} \mid x_{i+3}) + \cdots$$

In an inhomogeneous Markov chain of order n, the probability of each base is conditioned on the previous n bases, instead of just 1. For a second-order chain there are 64 different conditional probabilities corresponding to the 64 possible trinucleotides. For such a second order chain the score would be:

$$lp_1(x_i \mid x_{i-1}, x_{i-2}) + lp_2(x_{i+1} \mid x_i, x_{i-1}) + lp_3(x_{i+2} \mid x_{i+1}, x_i) + \cdots$$

A second-order chain spans a full codon, and usually that is the minimum order to use for scoring coding regions. Most gene finders use order 4 or 5, which correspond to pentamer or hexamer statistics.

Usually the same sensor is used for coding potential in the four different exon functions above, so they differ only in the signal sensors they use. The signal scores, such as splice site scores, are usually obtained from position-dependent score matrices (e.g. GeneParser and FGENEH) or neural networks (e.g. Genie). Often these 'signal sensors' are distinguished from the 'content sensors' in the dynamic programming algorithm (Figure 11.1). Although more natural in some formulations, it does not make any principal difference. Also, it does not make

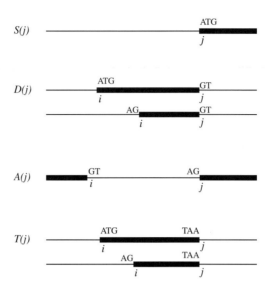

Figure 11.1   *Illustration of the dynamic programming algorithm described in the text. Potential coding regions are shown by a heavy line. Possible splice sites are marked with their consensus pattern (GT for donor and AG for acceptor), possible start codons are marked with ATG, and possible stop codons with TAA (one of the three stop codons).*

any difference if the score for splice sites is part of the intron score or the exon score.

Let us first consider the problem of combining coding regions optimally into exactly one complete gene without regard to the frame being consistent for the entire gene. It is assumed that the scores are additive, i.e. the optimal gene structure is the one with the largest score added up over all the components. Then a dynamic programming algorithm is as follows. For each position j in the sequence calculate:

1. Score for translation start at position j

$$S(j) = \texttt{intergenic}(1, j-1)$$

2. Score for best partial gene up to a donor at position j

$$D(j) = \max_{i < j} \begin{cases} S(i) + \texttt{initial\_exon}(i, j-1) \\ A(i) + \texttt{internal\_exon}(i, j-1) \end{cases}$$

3. Score for best partial gene up to an acceptor at position j

$$A(j) = \max_{i < j} [D(i) + \texttt{intron}(i, j-1)]$$

(j is the first base of the exon)

4. Score for best complete gene up to a stop codon at position j

$$T(j) = \max_{i < j} \begin{cases} A(i) + \texttt{terminal\_exon}(i,j-1) \\ S(i) + \texttt{single\_exon}(i,j-1) \end{cases}$$

(j is the base immediately after the stop codon which covers base $j-3$ to $j-1$).

See Figure 11.1 for a graphical representation of these terms. These numbers are calculated recursively from the beginning to the end of the sequence, $j = 1$, ..., L, and at the end

$$\max_{i < L} [T(i) + \texttt{intergenic}(i, L)]$$

yields the score of the optimal gene structure. To find that structure, one needs to save the i maximizing the expression at each step of the process as well as which of the terms maximized it (whether it was the term with $A(i)$ or $S(i)$ in the steps, where there is a choice). Then starting from the last step one can go backwards through each step and thus reconstruct the gene structure with the highest score.

As it stands, the computation time of the algorithm is at least proportional to the square of the length of the sequence. However, this can be reduced drastically by considering only valid signals; for example, calculate $S(j)$ only when there is a potential start codon (ATG) at position j; calculate D and A only at positions with potential splice sites; etc. When calculating exon score, one need not consider positions (i) further back than the first stop codon and, because open reading frames are rarely very long, this will also limit the calculation time significantly. These and other constraints make use of such an algorithm feasible even for long sequences.

This is essentially the algorithm used in GeneParser, except that the requirement of exactly one complete gene is relaxed, so that no genes or partial genes can also be predicted. Usually a genomic DNA sequence contains only parts of a gene or a mixture of whole genes and pieces of genes at the ends. Sometimes a cosmid from the human genome contains no genes or is part of a huge intron. It is easy to extend an algorithm like the one above to predict any number of genes and partial genes. For instance, to allow for one or more genes one need only change rule 1 to:

$$S(j) = \max_{i < j} \begin{cases} \texttt{intergenic}(1,j-1) \\ T(i) + \texttt{intergenic}(i,j-1) \end{cases}$$

This, of course, is still a fairly simplified algorithm. The second extension one should include is reading frame consistency between exons. This is fairly easy to do by having three different D values D_0, D_1 and D_2 and three different A values A_0, A_1 and A_2. The variables indexed with 0 are used for introns splicing between two codons; the ones indexed with 1 are used for those splicing after the first base in a codon, etc. Now the rules above need to be changed to inforce consistency, so, for instance, rule 3 would change to:

$$A_k(j) = \max_{i < j} [D_k(i) + \texttt{intron}(i, j - 1)]$$

Similar changes are needed for all the other rules that use A or D. Furthermore, the exon scoring functions need to be frame-aware, so one might have a score for each frame k, so $\texttt{internal\_exon}_k(\texttt{i,j})$ would mean an internal exon in which the first base corresponded to the kth codon position. Then rule 2 would become:

$$D_l(j) = \max_{i < j} \begin{cases} S(i) + \texttt{initial\_exon}(i, j - 1) & \text{if } l = (j - i) \bmod 3 \\ A_k(i) + \texttt{internal\_exon}_{k+1}(i, j - 1) & k = (j - i + l) \bmod 3 \end{cases}$$

where mod means modulus. The conditions on k and l are necessary for a consistent frame. Similarly, rule 4 would change to:

$$T(j) = \max_{i < j} \begin{cases} A_k(i) + \texttt{terminal\_exon}(i, j - 1) & \text{if } k = (j - i) \bmod 3 \\ S(i) + \texttt{single\_exon}(i, j - 1) \end{cases}$$

It is also possible to integrate sensors for promoters and the UTRs. This is done by adding new variables to the dynamic programming. For instance, one might have a variable I_j for transcription initiation. In this case the translation start depends on that variable; so, for instance, the $S(j)$ score in rule 1 would have to change to something like $S(j) = \max_{i < j}(I_j + \texttt{5'UTR\_score}(i, j - 1))$.

Now we have a method for combining sensors. There is, however, one problem that has not been addressed, which is how to weigh the output of the sensors against one another. Usually each sensor is estimated or trained independently of all the others, and they may use completely different scales. Similarly, if more than one measure of coding potential is used, it is likely that they are not equally reliable, and one would like to put a higher weight on the most reliable ones.

There are two ways of solving this problem: one is to use fully probabilistic models and the other is to train the weights in some way. The first of these methods will be discussed in the next section. The second method is that used in GeneParser, where several sensors are used for the various regions; there are, for instance, three different measures for coding potential (not counting database hits). The output of these sensors is combined by a neural network which is optimized to give the best predictions on a training set. In (Stormo and Haussler, 1994) a general method is given for optimizing the weights of the individual sensors.

3 STATE MODELS

The idea of combining the predictions into a complete gene structure is that

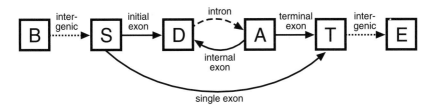

Figure 11.2   *A finite state automaton corresponding to the simple DP algorithm.*

the 'grammatical' constraints can rule out some wrong exon assemblies. The grammatical structure of the problem has been stressed by David Searls (Searls, 1992; Dong and Searls, 1994), who also proposed using the methods of formal grammars from computer science and linguistics. The dynamic programming can often be described conveniently by some sort of finite state automaton (Searls and Murphy, 1995; Durbin *et al.*, 1998). A model might have a state for translation start (S), one for donor sites (D), one for acceptor sites (A) and one for translation termination (T). Each time a transition is made from one state to another, a score (or a penalty) is added. For the transition from the donor state to the acceptor state, the intron score is added to the total score, and so on. In Figure 11.2 the state diagram is shown for the simple dynamic programming algorithm above. For each variable in the algorithm there is a corresponding state with the same name, and also a begin and end state is needed.

The advantage of such a formulation is that the dynamic programming for finding the maximum score (or minimum penalty) is of a more general type, and therefore adding new states or new transitions is easy. For instance, drawing the state diagram for a more general dynamic programming algorithm that allows for any number of genes and also partial genes is straightforward (Figure 11.3), whereas it is involved to write down. Similarly the state diagram for the frame-aware algorithm sketched out above is shown in Figure 11.4.

If the scores used are log probabilities or log odds, then a finite state automaton is essentially a hidden Markov model (HMM), and these have been introduced recently into gene finding by several groups. The only fundamental difference from the dynamic programming schemes discussed above is that these models are fully probabilistic, which has certain advantages. One of the advantages is that the weighting problem is easier.

VEIL (Henderson *et al.*, 1997) is an application of an HMM to the gene finding problem. In this model all the sensors are HMMs. The exon module is essentially a first-order inhomogeneous Markov chain, which is described above. This is the natural order for implementation in an HMM, because then each of the conditional probabilities of the inhomogeneous Markov chain corresponds

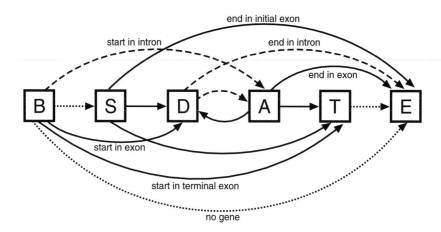

Figure 11.3   *The model of Figure 11.2 with transitions added that allow for prediction of any number of genes and partial genes where the sequence starts or ends in the middle of an exon or intron.*

to the probability of a transition from one state to the next in the HMM. It is not possible to avoid stop codons in the reading frame when using a first-order model, but in VEIL a few more states are added in a clever way, which makes the probability of a stop codon zero. Sensors for splice sites are made in a similar way. The individual modules are then combined essentially as in Figure 11.2 (i.e.

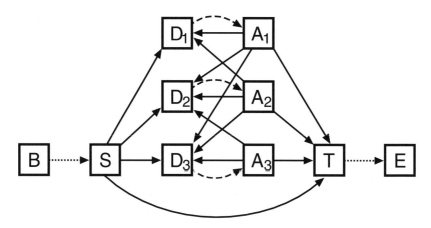

Figure 11.4   *A model that ensures frame consistency throughout a gene. As in the two previous figures, dotted lines correspond to intergenic regions, dashed to introns and full lines to coding regions (exons).*

frame consistency is not enforced). The combined model is one big HMM, and all the transitions have associated probabilities. These probabilities can be estimated from a set of training data by a maximum likelihood method. For combining the models this essentially boils down to counting occurrences of the different types of transitions in the dataset. Therefore, the implicit weighting of the individual sensors is not really an issue.

Although the way the optimal gene structure is found is similar in spirit to the dynamic programming above, it looks quite different in practice. This is because the dynamic programming is done at the level of the individual states in all the submodels; there are more than 200 such states in VEIL. Because the model is fully probabilistic, one can calculate the probability of any sequence of states for a given DNA sequence. This state sequence (called a path) determines the assignment of exons and introns. If the path goes through the exon model, that part of the sequence is labelled as exon; if it goes through the intron model it is labelled intron, and so forth. The dynamic programming algorithm, which is called the Viterbi algorithm, finds the *most probable* path through the model for a given sequence, and from this the predicted gene structure is derived (see Rabiner (1989) for a general introduction to HMMs).

This probabilistic model has the advantage of solving the problem of weighting the individual sensors. The maximum likelihood estimation of the parameters can be shown to be optimal if there are sufficient training data, and if the statistical nature of genes can be described by such a model. A weak part of VEIL is the first-order exon model, which is probably not capable of capturing the statistics of coding regions, and most other methods use fourth- or fifth-order models.

A HMM-based gene finder called HMMgene is currently being developed. The basic method is the same as VEIL, but it includes several extensions to the standard HMM methodology, which are described by Krogh (1997). One of the most important is that coding regions are modelled by a fourth-order inhomogeneous Markov chain instead of a first-order chain. This is done by an almost trivial extension of the standard HMM formalism, which allows a Markov chain of any order in a state of the model, whereas the standard HMM has a simple unconditional probability distribution over the four bases (corresponding to 0th order). The model is frame-aware and can predict any number of genes and partial genes, so the overall structure of the model is as in Figure 11.4 with transitions added to allow for begin and end in introns, as in Figure 11.3.

As already mentioned, the maximum likelihood estimation method works well if the model structure can describe the true statistics of genes. This is a very idealized assumption, and therefore HMMgene uses another method for estimating the parameters called conditional maximum likelihood (Juang and Rabiner, 1991; Krogh, 1994). Loosely speaking, maximum likelihood maximizes the probability of the DNA sequences in the training set, whereas conditional maximum likelihood maximizes the probability of the gene structures of these sequences, which, after all, is what we are interested in. This kind of optimization is conceptually similar to that used in GeneParser, where the prediction accuracy is also optimized. HMMgene also uses a dynamic programming algorithm different from the Viterbi algorithm for prediction of the gene structure. All of these methods have contributed to a high performance of HMMgene.

Genie is another example of a probabilistic state model which is called a generalized HMM (Kulp *et al.*, 1996; Reese *et al.*, 1997). Figure 11.4 is in fact Genie's state structure, and both this figure and Figure 11.2 are essentially copied from Kulp *et al.* (1996). In Genie, the signal sensors (splice sites) and content sensors (coding potential) are neural networks, and the output of these networks is interpreted as probabilities. This interpretation requires estimation of additional probability parameters which work like weights on the sensors. So, although it is formulated as a probabilistic model, the weighting problem still appears in disguise. The algorithm for prediction is almost identical to the dynamic programming algorithm of the last section. A version of Genie also includes database similarities as part of the exon sensor (Kulp *et al.*, 1997).

There are two main advantages of generalized HMMs compared with standard HMMs. First, the individual sensors can be of any type, such as neural networks, whereas in a standard HMM they are restricted by the HMM framework. Second, the length distribution (of, for example, coding regions) can be taken into account explicitly, whereas the natural length distribution for an HMM is a geometric distribution, which decays exponentially with the length. However, it is possible to have a fairly advanced length modelling in an HMM if several states are used (Durbin *et al.*, 1998). The advantage of a system like HMMgene, on the other hand, is that it is one integrated model, which can be optimized all at once for maximum prediction accuracy.

Another gene finder based on a generalized HMM is GENSCAN (Burge and Karlin, 1997). The main differences between the GENSCAN state structure and that of Genie or HMMgene is that GENSCAN models the sequence in both directions simultaneously. In many gene finders, such as those described above, genes are first predicted on one strand, and then on the other. Modelling both strands simultaneously was done very successfully in GeneMark, and a similar method is implemented in GENSCAN. One advantage (and perhaps the main one) is that this construction avoids predictions of overlapping genes on the two strands, which presumably are very rare in the human genome. GENSCAN models any number of genes and partial genes like HMMgene. The sensors in GENSCAN are similar to those used in HMMgene. For instance, the coding sensor is a fifth-order inhomogeneous Markov chain. The signal sensors are essentially position-dependent weight matrices, and thus are also very similar to those of HMMgene, but there are more advanced features in the splice site models. GENSCAN also model promoters and the 5' and 3' UTRs.

4 PERFORMANCE COMPARISON

Comparing gene finders is difficult. In Burset and Guigo (1996) several gene finders were compared on a dataset consisting of 570 mammalian sequences containing one complete gene with at least one intron. In Table 11.1 the performance on this dataset is shown for some of the gene finders mentioned here (see Chapter 10 for numbers on other gene finders). There are several problems with such a comparison. First, most of the gene finders were trained on data that had genes homologous to some of the sequences in this test set. As this overlap varies among the gene finders, the results are hard to compare. Second, whereas the initial test was done by an independent group, most of the numbers shown in the table are obtained by the developers themselves *after* the comparison was published. This makes it possible to keep tuning the method until it performs well on these particular data (only the values for FGENEH, GeneID and GeneParser are from the original study). Finally, it is likely that some of the gene finders in the comparison have been further developed since the test.

To test gene finders reliably they need to be trained and tested on the same datasets. This would require that all developers agree on a dataset and do cross-validation where the gene finder is trained on, for example, 9/10 of the

Table 11.1 Comparison of gene finders on the Burset–Guigo data

Performance in % Burset–Guigo data	Base		Exon			
	Sn	Sp	Sn	Sp	ME	WE
GeneParser2	66	79	35	40	29	17
GeneID	63	81	44	46	28	24
FGENEH	77	88	61	64	15	12
VEIL	83	72	53	49	19	NA
Genie	78	84	61	64	15	16
HMMgene	88	94	74	78	13	8
GENSCAN	93	93	78	81	9	5

These gene finders do not use database searches. Sensitivity (Sn) and specificity (Sp) are shown at both the single nucleotide and exact exon level. The last two columns show missing exons (ME) (the percentage of true exons that do not overlap with a predicted one) and wrong exons (WE) (the number of predicted ones that do not overlap a correct one). See Chapter 10 for a discussion of these performance measures. The values for GeneParser2, GeneID and FGENEH are taken from the original comparison (Burset and Guigo, 1996), in which FGENEH showed the best performance. Values for Genie are from Reese et al. (1997), those for HMMgene are from Krogh (1997), and those for GENSCAN are from Burge and Karlin (1997). Note that the results for VEIL are for five-fold cross-validation on the Burset–Guigo data.

data and tested on the remaining 1/10. This process is repeated 10 times until all the data have been used for testing and the performance measures are averaged. As a start in that direction the Genie group has made its dataset of almost 400 human genes available for other researchers, and in fact Genie and HMMgene have been cross-validated on this set (see Reese *et al.* (1997) and Krogh (1997) for results). In the Genie data significant homologies between genes have been removed, and it is restricted to sequences with exactly one complete gene. The performance is generally lower on these data than on the Burset–Guigo data, first, because the training set is non-homologous with the test set because of cross-validation, and second, because the sequences are longer and seem to be generally harder. Hopefully in the future there will be good datasets of reliably annotated genomic sequence for the gene finders to be trained and tested on.

Both Genie and HMMgene were trained on the Genie dataset, and therefore the results shown in Table 11.1 are comparable. GENSCAN also used this dataset for training, but additionally used a set of almost 2000 complete human cDNA sequences. This much larger dataset allows a reliable estimation of the fifth order coding model used, whereas my experiments with HMMgene showed that using a coding model of order higher than 4 based on the Genie data alone was not successful. Furthermore, it is almost certain that there are more homologies between GENSCAN's dataset and the test set. GENSCAN stands out as the best in Table 11.1, and there is no doubt that GENSCAN is currently among the best gene finders. However, direct comparison based on these numbers is impossible.

5   CONCLUSION

In this paper I have discussed ways of combining information of splice sites, coding potential, etc. into a prediction of a complete gene structure. It is quite easy to extend most of these methods to include sensors for promoters, 5' and 3' untranslated regions, poly(A)site, and so forth. In HMMgene this has given only marginal improvements, and is currently not included. As more knowledge is gained about gene structure and regulation, it is quite possible that incorporation of more sensors can significantly improve prediction.

Most current gene finders are developed for DNA sequences with no or very few errors and for 'textbook genes' that all have the GT/AG consensus splice sites, no frame shifts, no alternative splicing, and so forth. Many of these complications can be dealt with in most of the models mentioned here, but no doubt at the cost of worse performance. Although some of the problems have been taken up already, these issues are mainly topics of future research.

HMMgene, which was described briefly, is still being developed and the values shown have improved somewhat. It is possible to follow the development and test the current version at web site http://www.cbs.dtu.dk/services/HMMgene/. Links to some of the other gene finders mentioned can be found on my web page http://www.cbs.dtu.dk/krogh/genefinding.html.

ACKNOWLEDGEMENTS

Several discussions with Richard Durbin about dynamic programming and finite state machines are gratefully acknowledged. I would like to thank David Ussery and Nikolaj Blom for comments on the manuscript. My work is supported by the Danish National Research Foundation.

REFERENCES

Birney E and Durbin R (1997) Dynamite: a flexible code generating language for dynamic programming methods used in sequence comparison. In *Proceedings of the Fifth International Conference on Intelligent Systems for Molecular Biology*, pp. 56–64 (eds T Gaasterland *et al.*), AAAI Press, Menlo Park, CA.

Borodovsky M and McIninch J (1993) GENMARK: parallel gene recognition for both DNA strands. *Computers and Chemistry* 17, 123–133.

Brunak S, Engelbrecht J and Knudsen S (1991) Prediction of human mRNA donor and acceptor sites from the DNA sequence. *J. Mol. Biol.* 220, 49–65.

Burge C and Karlin S (1997) Prediction of complete gene structure in human genomic DNA. *J. Mol. Biol.* 268, 78–94.

Burset M and Guigo R (1996) Evaluation of gene structure prediction programs. *Genomics* 34, 353–367.

Dong S and Searls DB (1994) Gene structure prediction by linguistic methods. *Genomics* 23, 540–551.

Durbin RM, Eddy SR, Krogh A and Mitchison G (1998) *Biological Sequence Analysis*. Cambridge University Press, Cambridge.

Fields CA and Soderlund CA (1990) gm: A practical tool for automating DNA sequence analysis. *Comput. Appl. Biosci.* 6, 263–270.

Gelfand MS and Roytberg MA (1993) Prediction of the exon–intron structure by a dynamic programming approach. *Biosystems* 30, 173–182.

Gelfand MS, Mironov AA and Pevzner PA (1996) Gene recognition via spliced sequence alignment. *Proc. Natl. Acad. Sci. U.S.A.* 93, 9061–9066.

Guigo R, Knudsen S, Drake N and Smith T (1992) Prediction of gene structure. *J. Mol. Biol.* 226, 141–157.

Henderson J, Salzberg S and Fasman KH (1997) Finding genes in DNA with a hidden Markov model. *J. Comput. Biol.* 4, 127–141.

Juang BH and Rabiner LR (1991) Hidden Markov models for speech recognition. *Technometrics* **33**, 251–272.

Krogh A (1994) Hidden Markov models for labeled sequences. In *Proceedings of the 12th IAPR International Conference on Pattern Recognition*, pp. 140–144. IEEE Computer Society Press, Los Alamitos, CA.

Krogh A (1997) Two methods for improving performance of a HMM and their application for gene finding. In *Proceedings of the Fifth International Conference on Intelligent Systems for Molecular Biology*, pp. 179–186 (eds Gaasterland *et al.*), AAAI Press, Menlo Park, CA.

Kulp, D, Haussler D, Reese MG and Eeckman FH (1996) A generalized hidden Markov model for the recognition of human genes in DNA. In *Proceedings of the Fourth Conference on Intelligent Systems in Molecular Biology*, pp. 134–142. (eds DJ States *et al.*), AAAI Press, Menlo Park, CA.

Kulp D, Haussler D, Reese MG and Eeckman FH (1997) Integrating database homology in a probabilistic gene structure model. In *Proceedings of the Pacific Symposium on Biocomputing* (eds RB Altman, AK Dunker, L Hunter and TE Klein) World Scientific, New York.

Rabiner LR (1989) A tutorial on hidden Markov models and selected applications in speech recognition. *Proc. IEEE* **77**, 257–286.

Reese MG, Eeckman FH, Kulp D and Haussler D (1997) Improved splice site detection in Genie. *J. Comput. Biol.*, **4**, 311–323.

Searls DB (1992) The linguistics of DNA. *Am. Sci.* 80, 579–591.

Searls DB and Murphy KP (1995) Automata–theoretic models of mutation and alignment. In *Proceedings of the Third International Conference on Intelligent Systems for Molecular Biology*, Vol. 3, pp. 341–349 (eds C Rawlings *et al.*), AAAI Press, Menlo Park, CA.

Snyder EE and Stormo GD (1993) Identification of coding regions in genomic DNA sequences: an application of dynamic programming and neural networks. *Nucleic Acids Res.* **21**, 607–613.

Snyder EE and Stormo GD (1995) Identification of protein coding regions in genomic DNA. *J. Mol. Biol.* **248**, 1–18.

Solovyev VV, Salamov AA and Lawrence CB (1994) Predicting internal exons by oligonucleotide composition and discriminant analysis of spliceable open reading frames. *Nucleic Acids Res.* **22**, 5156–5163.

Solovyev VV, Salamov AA and Lawrence CB (1995) Identification of human gene structure using linear discriminant functions and dynamic programming. In *Proceedings of the Third International Conference on Intelligent Systems for Molecular Biology*, Vol. 3, pp. 367–375 (eds C Rawlings *et al.*) AAAI Press, Menlo Park, CA.

Stormo GD and Haussler D (1994) Optimally parsing a sequence into different classes based on multiple types of evidence. In *Proceedings of the Second International Conference on Intelligent Systems for Molecular Biology*, pp. 369–375.

Wu TD (1996) A segment-based dynamic programming algorithm for predicting gene structure. *J. Comput. Biol.* **3**, 375–394.

Xu Y, Einstein JR, Mural RJ, Shah M and Uberbacher EC (1994) An improved system for exon recognition and gene modeling in human DNA sequences. In *Proceedings of the Second International Conference on Intelligent Systems for Molecular Biology*, pp. 376–384.

12   Gene-Expression Databases

Duncan Davidson[1], Martin Ringwald[2] and Christophe Dubreuil[1]

[1]Developmental Genetics Section, MRC Human Genetics Unit, Western General Hospital, Crewe Road, Edinburgh EH4 2XU, UK and [2]The Jackson Laboratory, 600 Main Street, Bar Harbor, ME 04609, USA

1   INTRODUCTION

The Human Genome Mapping Project and other sequencing projects are rapidly identifying new genes and expressed sequences. In the wake of these projects will follow a much larger and more challenging endeavour, to characterize the functions of these genes and to gain an understanding of the networks in which they interact. Studies of gene function at the cell and tissue level rely primarily on evidence from the effects of mutation or of modified expressions, and on gene-expression assays. Expression patterns point the way to experimental investigations of particular cell compartments or particular organs. Moreover, with growing number of expression patterns being reported it is becoming increasingly useful to scan the data for genes with overlapping or complementary expression in order to frame novel hypotheses about genetic interactions which might then be tested by molecular genetic experiments.

In response to the very large amounts of gene-expression data that must be accessed and compared in these functional studies a variety of gene-expression databases have been set up (Table 12.1); most are in the early stages of development and the list will undoubtedly expand. Many of these databases have been initiated, and are presently run, by biologists with their own active programmes of research. They are often, therefore, a direct and simple response to the immediate requirements described above. This simplicity is paralleled by the relative paucity of data presently contained in the databases but contrasts starkly with the complexity and quantity of information – potentially

Guide to Human Genome Computing, 2nd edition   Copyright © 1998 Academic Press Limited
ISBN 0-12-102051-7   All rights of reproduction in any form reserved

Table 12.1   Internet addresses of gene-expression and related databases

Database	Internet address
ACeDB	http://www.sanger.ac.uk/
Flyview	http://flyview.uni-muenster.de/
Flybrain Project	http://flybrain.uni-freiburg.de/
Flybase	http://flybase.bio.indiana.edu:82/
Zebrafish Database	http://zfish.uoregon.edu/
Xenopus Molecular Marker Resource	http://vize222.zo.utexas.edu/
Mouse Gene Expression Information Resource	http://genex.hgu.mrc.ac.uk/ http://www.informatics.jax.org/
TBASE	http://www.gdb.org/tbase/tbase.html
dbEST	http://www.ncbi.nlm.nih.gov/
TIGR Database	http://www.tigr.org/tdb/tdb.html
Kidney Development Database	http://www.ana.ed.ac.uk/anatomy/kidbase/kidhome.html
Organogenesis Database	http://www.ana.ed.ac.uk/anatomy/kidbase/orghome.html
Tooth Gene Expression Database	http://honeybee.helsinki.fi/toothexp/
Ion Channel Database	http://parrot.le.ac.uk/csn/
Human Anatomy Database	http://www.ana.ed.ac.uk/anatomy/database/humat/

much greater than for sequence data – that remains to be recorded in gene-expression databases.

One problem in writing about gene-expression databases is that, for the few that are currently available, operation is simple and self-explanatory, whereas the more complex databases that are being developed are not, as yet, operational. In this chapter, therefore, we aim to give a brief comparative overview of present databases and to include critical assessments of issues that arise from the nature of the data and that must be addressed in the future. We begin by outlining the scope of the current databases then describe the relevant classes of data, and briefly describe the databases that are available or are being developed. Lastly, we discuss future directions for the application of gene-expression databases beyond their present functions in the publication, storage and scanning of data.

2   THE SCOPE OF GENE-EXPRESSION DATABASES

Gene-expression databases are available, or are under development, for most of the model organisms studied by developmental biologists and may, in the future, also be available for human development (see Table 12.1 for the Internet addresses of databases referred to in this paper; see also Davidson *et al.*, (1988) for a review). Some databases apply only to particular organs (e.g. the Tooth Database and the Kidney Development Database), others to particular processes (e.g. the Ion Channel Database); a few have a much wider range, being designed to document gene-expression in cells *in vitro*, in embryos and, potentially, in the adult organism, for example the Zebrafish Database and the Mouse Gene-Expression Information Resource (MGEIR). Most deal only with organisms of the wild type or of inbred strains, but all have the potential to include data from mutants and disease states. One, TBASE, concentrates on transgenic and mutant organisms (currently *Drosophila*, mouse, rat and pig).

The scale of these databases varies widely. Some are concerned only with gene-expression. Most of these simply list genes that are expressed in particular structures at approximate times during development or in the adult. If the user wishes further information regarding the pattern of expression (for instance, whether it is uniform or graded, weak or strong), or descriptions of methods, contact addresses, etc., these must be obtained from the original publication which is usually cited in the database entry. Some of these relatively simple databases do, however, contain images of *in situ* gene-expression data, which provide limited, but useful, views of the actual pattern. In the future, databases will be used in a more investigative way, and for this purpose several gene-expression databases that are now being developed will record all the data necessary for making investigative queries and will be linked to other databases containing comprehensive information on genomic sequence, gene mapping information and bibliographical data. Flyview and the MGEIR are examples of these more complex databases.

For some databases, gene-expression data is only a small part of the information held. A good example of this type of database is ACeDB (A *C. elegans* Database) which contains genome sequence, genetic and physical mapping data; another example is the Zebrafish Database which will contain a wide range of phenotypic and genetic data as well as contact addresses, and so on. Expression databases that are linked to, or contain, sequence or phenotypic data will be a rich resource for gene function studies and will support questions such as, 'Which genes that map to a particular chromosome site are expressed in a particular tissue?' or 'Which genes are expressed in a particular site and have mutant alleles that affect a particular process?'

The query and search capabilities of most currently available gene-expression databases are primitive. In the future, it is clear that queries will be required to meet combinations of constraints, for example, 'Which genes are expressed in the apical ectoderm of the limb, but not in the subadjacent mesenchyme?'. Equally, it is clear that report-making capabilities will be required, with the presentation of search results in tabular form, or with the combination of several images of spatially mapped data. The ability to compare whole patterns of gene-expression will also be important, for example to address the question, 'Which genes are expressed in the regions where genes X and Y are coexpressed?' or 'Which genes have the same pattern of expression as gene X in the limb?' Spatial and temporal aspects of queries are likely to become increasingly important, for example, 'Which genes show a complementary expression to gene X in the developing telencephalon?' or 'Which genes are expressed after or during, but not before, the expression of *MyoD* during muscle development?' It will also be important to gain access to information about which splice forms may be detected in a gene-expression experiment so that one can ask, for example, 'Which genes are coexpressed with one but not another splice form of a given transcription factor?' The inclusion of gene-expression data from mutant organisms or disease states will also be important. Indeed, one might foresee disease model databases in which gene-expression and other data in diseased tissue and under different theraputic regimens would be shared within particular interest groups in the biomedical field.

Moving forward from the results of a query, several of the existing databases provide information that will be useful for the further investigation of gene function. Most provide references to papers that may contain more complete descriptions of results and methods, as well as pointers to genetic and biochemical evidence relevant to gene function. Some, for example Flyview and the *Xenopus* Molecular Marker Resource, provide contact addresses and sources of probes or enhancer-trap lines.

3   A SURVEY OF GENE-EXPRESSION DATA

Although databases vary in their approach to recording gene-expression information, a comprehensive dataset can be identified of which subsets are documented in different databases. In this section we consider this comprehensive dataset and discuss selected issues relating to it. Few of the present databases record all of this information. Section 4 describes individual databases.

3.1   Gene Name

Most databases use standard gene names that are recognized by the appropriate nomenclature committees. The use of standard text for gene names and symbols is important for searching and interpreting the data and provides a key link for relating information in different databases.

3.2 Method

Although few gene-expression databases contain information on the methods by which data for each entry were obtained, this information may be crucial in the interpretation of results. Thus, while databases generally indicate whether expression was assayed at the RNA or protein level, most do not describe the probe or antibody used, making it impossible to assess the specificity of the assay or which splice form or gene product has been detected. The approach used by the GXD and MGEIR (see below) is to specify the probe in sufficient detail to enable this information to be matched to up-to-date sequence data and information on alternative gene products at the time of query, so that the user could assess which gene products were likely to be detected in the assay. This exemplifies the general point that, to be of maximum use in investigative searches, information in the database should relate to the original data rather than to an interpretation of the results.

To assess the sensitivity and resolution of the assay, it is also important to record whether gene-expression was assayed *in situ* (by *in situ* hybridization, immunohistochemistry or histochemistry) or in homogenized material (e.g. data from RNase-protection, reverse transcriptase–polymerase chain reaction (RT-PCR), gels, blots, cDNA libraries, solid state 'chip' technology, etc.). Information on whether *in situ* assays were carried out in sectioned material or in whole-mount preparations is also important as this affects the expected resolution of the data: it is generally more difficult to determine which tissues display gene-expression in whole-mount preparations than in sectioned material. Critical aspects of the assay, such as hybridization and wash temperatures, salt concentrations and antibody blocking agents, may also be necessary to interpret the specificity of the data, and those compiling databases should be encouraged to include this information as an annotation if the author of the data decides that this is important for interpreting the results of the experiment.

It is clearly also relevant to record tissues that were examined but in which no expression was detected, as distinct from tissues or stages of development that were not examined. Currently most databases do not include this information.

3.3 Temporal Aspects of Gene-Expression

Like standard gene names, standard developmental stages provide a potentially important link to other databases. For many organisms, standard stages of development from fertilization to the adult have been described and most databases use these, although sometimes rather loosely. One particular difficulty is that individual embryos of the same age may vary considerably in developmental stage. Yet by far the most common temporal description is in terms of time elapsed since fertilization. Where the database permits, developmental data should be recorded in relation to the standard stages. For the mouse, for example, the stages described by Theiler (1989) are preferable to 'days post-conception (d.p.c.)'.

A further problem is that dynamic aspects of gene-expression may call for a finer subdivision of time than the standard developmental stages. In particular, it may be appropriate to describe the expression of a gene in relation to the time of key molecular or cellular events such as the expression of a developmentally significant marker gene or a change in environmental conditions. This is particularly relevant to interpreting the expression of genes involved in developmental processes (such as muscle cell differentiation) which are more or less advanced in different regions of the same embryo. None of the databases currently provides this form of data. This could be done, for example, by recording expression as 'before, concurrent with, or after' the onset of expression of a marker gene as assayed in the same experiment.

3.4   Spatial Aspects of Gene-Expression

Spatial patterns of gene-expression can be recorded in text or images. Some currently available databases that store gene-expression information (e.g. the *Xenopus* Molecular Marker Resource and TBASE) use free text descriptions, whereas others (e.g. Flyview) use controlled vocabularies to describe sites of gene-expression. Controlled vocabularies include standard anatomical nomenclature systems and can be used consistently in different databases relating to the same organism. For *Caenorhabditis elegans* there is a unique means of identifying individual cells on the basis of lineage. A controlled vocabulary for *Drosophila* is given in Flybase; indeed, a subset of this vocabulary describes subcellular components and could form the basis for subcellular descriptions in other species. The MGEIR will include an anatomical description of the mouse embryo which is organized in an open-ended database to which finer detail can be added as required. The mouse anatomical nomenclature is being used, where appropriate, to help formulate descriptions of zebrafish and human embryos. It is to be hoped that, ultimately, the use of standard nomenclature systems will not only simplify the use of each of these databases, but will enable cross-database links.

A novel non-textual approach to recording gene-expression information is employed by the Tooth Database, in which gene-expression is assigned to tissues that are represented in diagrams of tooth primordia at successive stages of development. The use of diagrams removes the need for the user to identify the tissue by name and reduces confusion over the use of different names. Data cannot, however, be assigned to part of a defined tissue.

In addition to textual records, many databases hold images that illustrate the details of complex gene-expression patterns. A major distinction here is between databases that document the expression of different genes in a series of independent images and those in which different gene-expression patterns are spatially mapped onto a single set of standard reference images. Spatial mapping allows immediate comparison between different patterns and, in principle, allows purely spatial searches. The main advantage of spatially mapped data over textual descriptions is that many gene-expression patterns do not map 1 : 1 with an anatomical description, so that gene-expression data often cannot be translated simply into a list of anatomical terms. This is particularly true of genes involved in mechanisms that operate in a spatial, rather than tissue-specific, context during the early stages of development when the anatomy of the organism is being established. The spatial mapping approach has considerable potential because it can link gene-expression data to other spatially distributed information of relevance to gene function, for example cell lineage, cell proliferation and cell death. A disadvantage is that the data are transformed from their original form. This is more laborious than simply cataloguing original images; moreover, the mapping must be done carefully to ensure that significant features are not lost or spurious features introduced. The *Drosophila* database being developed at the University of California by Hartenstein and colleagues (Hartenstein *et al.*, 1996), the Zebrafish Database and the MGEIR will employ spatial mapping.

As more gene-expression patterns are entered into databases it is likely that some will become established as reference patterns. These will be patterns that are clearly defined, invariant and dynamically stable, and for which probes or antibodies are readily available. These gene-expression domains will define a 'molecular anatomy' against which the expression of other genes can be described. Such a molecular anatomy may reflect a developmentally more profound subdivision of the embryo than the morphological subdivision that is presently recognized by classical anatomy.

3.5   Quantitative Data

In general, techniques for assaying gene-expression *in situ* are not quantitative, except in the sense that semiquantitative comparisons can be made within the same whole-mount preparation or histological section, and the best that can be done here is a simple 'weak/moderate/strong' annotation of the *in situ* signal or the inclusion of representative images to illustrate, for example, gradients of expression. Some planned databases (e.g. the MGEIR) will record textual information on semi-quantitative aspects of gene-expression.

Methods applied to homogenized material, such as Northern and Western blots and RT-PCR, can potentially provide more quantitative information, but comparisons are only possible within individual datasets which are usually rather small. Recently developed high-throughput expression methods, for example the analysis of high density cDNA arrays with complex probes (Nguyen *et al.*, 1995; Schena *et al.*, 1995; Piétu *et al.*, 1996; Lockhart *et al.*, 1996), are about to change this situation. They will generate very large datasets within which quantitative comparisons can be made. For the yeast, a large dataset is already publicly available (http://cmgm.stanford.edu/pbrown/explore/index.html). For the mouse, GXD and the MGEIR plan to store this type of data and to integrate them with other types of expression information.

3.6   Experimental Evidence Concerning Gene Products

Information on where and when a gene is expressed may be associated with other data on the characteristics of the gene product. For example, evidence of gene-expression in tissue extracts assayed electrophoretically may provide additional information on product size or post-translational processing. Few gene-expression databases currently store such data, and it is planned that the MGEIR, for example, should do so.

3.7   Data Submission

Each database publishes its own instructions for data submission on the World Wide Web (WWW), and these instructions are being updated continuously. Databases that provide on-line data submission by individual users should be distinguished from those in which data are entered centrally by the database team. At present, most operational gene-expression databases acquire data from the literature or, to a lesser extent, from experimental results sent by individual laboratories by ad hoc mailing. Data from these sources is entered into the database by database editors. Although annotation of expression data from the literature is important, it is clear that in the future the preferred mode of data acquisition for larger databases will be by electronic submission. This is the only practical way to deal with the rapidly increasing volume of data that is being generated. Furthermore, standard publications contain only a limited amount of expression data. For example, most of the raw data from *in situ* hybridization and immunohistochemistry studies are never published, but merely referred to in brief text summaries. For these reasons, databases like the Zebrafish Database and the MGEIR are developing electronic submission systems that will enable the acquisition of comprehensive and standardized expression data.

Issues of quality control arise for submissions other than for data that are abstracted from peer-reviewed journals. At present, most databases do not apply strict *quality* control, although the MGEIR plans to request that *in situ* data are accompanied by at least one image of original data to enable quality to be assessed. Generally, the *quantity* of data is not a limiting factor. A good gene-expression database can hold much more extensive and detailed information than can be published in a conventional paper. In particular circumstances, however, each entry may comprise only a small amount of data. For example, a single entry of data from an extensive *in situ* hybridization screen may include information from only a single section assayed using an uncharacterized probe. Such data would be useful and should therefore be acceptable provided the data quality is adequate and that the probe is made available to those for whom the effort of further characterization would be worthwhile.

An understanding of how data were aquired and entered into the database is important to the interpretation of any query. It is also important to realize that most databases are far from comprehensive and that much of the data is incomplete and may be inaccurate or entered only at a low resolution. In this light, gene-expression databases should be seen not as repositories of truth, but as tools by which to frame hypotheses which can then be tested by experiment.

3.8 Third-Party Annotation of Existing Data

None of the currently available gene-expression databases provides for annotations of the data by other users.

3.9 Linked Entries Within a Database

At present, gene-expression databases do not provide the possibility of grouping different entries, for example linking data on the expression of two genes to show that these were assayed in consecutive sections in the same experiment. Such information is potentially useful when comparing two genes or two products of the same gene, and may also be important to describe gene-expression by comparison with data on controls. In fact, none of the currently available databases deals explicitly with control data. At least some of the databases that are now being developed (e.g. the MGEIR) plan to address these issues.

3.10 Links to Other Databases

There is considerable potential for links between different gene-expression databases relating to the same organism, for example in the mouse between the Tooth Database, the Kidney Development Database and the MGEIR.

Links between gene-expression databases dealing with different species will also be important. This is particularly the case between mouse and human (Davidson and Baldock, 1997), but links between mouse, zebrafish and *Drosophila* databases will be important because genetic studies in fish and flies are providing an important starting point for identifying mammalian genes.

As discussed briefly in Section 2, links from gene-expression databases to databases dealing with sequence, genetic linkage, phenotypic and mutant data will be important in using evidence concerning gene-expression in studies of gene function. An additional way to link related information from different databases is to abstract data and combine them in a single resource aimed at a particular audience. An excellent example of this approach is the Ion Channel Database (Table 12.1).

All these links depend on the compatibility of data (particularly on the use of standard nomenclature for genes and anatomical components), the feasibility of relating data models, the practicality of inter-database links at the software level, and the motivation across the community to fund and implement the necessary work. At present, none of the gene-expression databases is linked in this way, although it is clear that the need for integration is well appreciated. Given the considerable costs of identifying genes and characterizing their expression, the effort required to ensure that these data can be used investigatively by linking the relevant databases is likely to be cost-effective.

4   A SURVEY OF GENE-EXPRESSION DATABASES

4.1   Access to the Databases

Access to the databases described here is straightforward as all are, or will be, available on the Internet and particularly on the WWW. The minimum requirement is, therefore, a computer with an adequate connection to the Internet and a WWW browser (see Chapter 1). In some cases, database access is achieved through simple web browsing (e.g. the Kidney Database and the *Xenopus* Molecular Marker Resource), but most already make use of newer features of the WWW, such as forms for queries or frames/tables for better displays, and these will require up-to-date browsers. Several databases currently, or in the near future, will also provide improved data interfaces using Java applets. These are small programs that run inside the web browser and their use requires a browser that can interpret Java. ACeDB provides Jade, a Java interface to the database. The Zebrafish Database will use Java to annotate images and display sequence data. The MGEIR will use Java to display the anatomy nomenclature in the

Mouse Embryo Anatomy Database and to interact with images in the WWW version of the Mouse Atlas.

4.2 Databases

The following information was up to date in July 1997. Further details of some of these databases are described in Bard (1997).

4.2.1 A C. elegans *Database (ACeDB)*

ACeDB stores gene-expression data as a part of a much wider range of information about *C. elegans*, in particular genetic and physical mapping data (clones and contigs), and the complete DNA sequence. Indeed, ACeDB is designed to integrate any form of experimental data in a common, easy-to-use format. ACeDB is available to authorized sites via the WWW: the database administrators release version code for Sun, Solaris, DEC(OSF) and SGI (IRIX) machines, and there is a Mac version, MACACE. The main interface is for Unix computers and uses an X-windows-based, mouse-driven, click-and-point navigation method. Users can obtain copies of the database for use on their own computers, to which they can add their own data. For the public database, data are submitted by users to be entered by the database curator. The database accepts both textual and original image data via e-mail or ftp. Text data are submitted as ASCII files that are read into the database in a standard tree-form structure. Images are added to a picture library and can be called from the database and displayed in a separate (xv) viewer (Unix versions only). The system provides query and table-making functions, bibliography searches, and general search engines. There are numerous display capabilities including configurable genetic map displays, physical map displays, and sequence feature displays for DNA. Gene-expression data can be searched by text string, or accessed through searches on the other types of data, including individual cells, cell groups, sequences, loci, clones and bibliographical information. This database is fully operational. Currently, most of the gene-expression data comes from just two laboratories and is not comprehensive. Submission by other workers is being encouraged.

Three databases exist, or are being developed, to store gene-expression data relating to *Drosophila* development (see 4.2.2–4.2.4).

4.2.2 Flyview

This database stores gene-expression data for enhancer-trap lines and some cloned genes, and is accessible via the WWW. It includes both text and images of original data, as well as bibliographical data. Textual descriptions are in free text, but using, as far as possible, the controlled vocabulary of Flybase. Published and unpublished data are accepted. Individual researchers are responsible for their own data, but the entries are made by the Flyview team. Data can be submitted via the WWW, by e-mail or ftp, and images can be submitted digitally, as slides or photographic prints, and even as original data (microscope slides). Most of the database records are linked to FlyBase and/or to the Encyclopaedia of *Drosophila*. The database can be browsed on the WWW and searched by text description using the controlled vocabulary of FlyBase. The database is operational. The enhancer-trap lines are available to the community and the database contains an index of stocks.

4.2.3   Database of Hartenstein et al. (1996)

This database aims to map gene-expression and other data spatially to digital models of the *Drosophila* embryo. The system will be available via the WWW and the software currently being used runs on Macintosh computers. Each model is being built from differential–interference–contrast (DIC) images of optical section planes taken at $2\text{-}\mu$m intervals through embryos. All the major organs will be outlined on these images, including the tissues visible under DIC optics (epidermis, musculature, central nervous system, intestinal tract) and cells that have been identified using specific markers. Using the outlined structures as landmarks, users will be able to enter data on labelled cells, gene-expression, etc. into these images of sections. The data will then be visualized in the context of the standard three-dimensional (3D) embryo model. The system is not yet fully operational.

4.2.4   The Flybrain Database

The database that is being produced as part of the Flybrain Project aims to provide information relating specifically to the structure and function of the *Drosophila* nervous system (Heisenberg and Kaiser, 1995). Gene-expression data will be only a part of much wider information which will include an atlas of the nervous system. This atlas aims to give an overview of how the nervous system is constructed, based on schematic representations, serial sections and informative images, each related to a common coordinate system. This atlas will provide the context, via hypertext links, for data that potentially include patterns of expression of genes and regulatory elements, as well as anatomical structures and genetic variants. Most of the data will relate to *D. melanogaster*, but information from other *Drosophila* species, and other *Diptera*, may be included for comparison. It is planned to link the database to FlyBase. This database is operational, but has very little gene-expression data.

4.2.5   Zebrafish Database

This will store gene-expression information as part of a much more comprehensive database that will include genetic and developmental data. Access will be via the WWW. In addition to gene-expression information, the database will have an atlas for staging embryos, an anatomical atlas of the adult (including standard anatomical nomenclature), information on mutants (including images), genetic information (gene names, cDNA and genomic sequence, markers, etc.), information on probes and antibodies, a bibliography, and lists of people working on zebrafish. The database will store gene-expression data in textual and image form. Text data will use a controlled vocabulary. Unpublished as well as published data will be accepted, but unpublished information will be marked so that users can assess the confidence level of each entry. Data will be entered by 'authorized users' and by the database team. The database will be accessible through a graphical interface to the WWW and 'guest' users will be able to run common types of searches directly from the WWW interface; other queries will be possible via the database team. It is planned to have a form of spatial searching based on graphical representations of the embryo. This database is not yet operational. It is planned, eventually, to link this database to those for other organisms, including the MGEIR.

4.2.6   Xenopus *Molecular Marker Resource (XMMR)*

This database lists genes and markers and documents their expression in the *Xenopus* embryo, linking this information to other genetic and developmental data. The database is available via the WWW. Entries include bibliographical references linked to Medline, notes on the availability of probes, etc., and links to sequence information. Gene-expression is recorded as free text and in images of whole-mount embryos and sections. Images can be submitted electronically or as hard copy, including photographs and projection slides. Entries to the database are made up as web pages and, for those who are not familiar with the methods of building a web page, help is available from Peter Vize, who runs the database. Text-string searches can be made across the text-content of the database, but, with the present volume of data available, it may be best to begin by scanning the index visually. Here, genes are listed according to the anatomical structures in which they are expressed; clicking on a gene in the list calls up pictures of gene-expression and a brief text description with references, and links to sequence databases, and addresses for available probes, antibodies, etc. This database is operational. The XMMR home page has links to a wide range of information on *Xenopus*, including standard stages, and lists of libraries and expression plasmids.

4.2.7   *Mouse Gene-Expression Information Resource (MGEIR)*

The MGEIR will be an integrated system comprising several different databases and will document gene-expression in the mouse embryo. The resource will be available via the WWW. Although the prototype versions of the data entry inter-face to the GXD part of the database (see below) currently run on Macintosh computers, it is intended that the integrated MGEIR will be accessible via platform-independent, textual and graphical interfaces. This database will be fully integrated with the Mouse Genome Database (MGD) (Baldock *et al.*, 1992; Ringwald *et al.*, 1994; Davidson *et al.*, 1998). The MGEIR will comprise: the GXD Gene-Expression Database, the Mouse Graphical Gene-Expression Database (GGED), the Mouse Atlas, and the Mouse Embryo Anatomy Database. Data will include patterns from *in situ* hybridization, histochemistry and immunohistochemistry, and information from assays on homogenized tissues. Gene-expression will be documented as text using the controlled vocabulary of the Mouse Embryo Anatomy Database, as images of original data, and as graph-ical data spatially mapped to the embryo models of the 3D Mouse Atlas. Retrospective data will be entered by the database team. New data will be sub-mitted by individual researchers and checked by the database team. Published and unpublished data will be marked accordingly and it is planned to offer, as an additional alternative, a peer review of submissions in conjunction with one or more established journals so that appropriate entries can achieve full publi-cation status. All the data will be searchable by textual and spatial searches. The database is not yet operational.

The individual parts of the MGEIR are as follows.

4.2.7.1   Mouse Embryo Anatomy Database

This database will contain the names of morphologically recognizable structures organized in a spatial hierarchy for each of the 26 developmental stages defined by Theiler (1989). Users will be able to browse the database via a Java interface. The data will include a thesaurus of alternative names, notes and references on contentious aspects of the nomenclature, information on tissue type/architecture, and data on tissue derivation and cell lineage (where reliable information exists), to enable the user to trace the origins and future develop-ment of each component. Comments on nomenclature and suggested additions can be addressed to j.bard@ed.ac.uk.

4.2.7.2   Mouse Atlas

This atlas of embryonic development will comprise digital 3D models of mouse embryos at successive stages of development. The model embryos, together with utility software, will be available as a set of CD-ROMs and via the WWW through a Java interface. The models will represent stages from fertilization to stage 22 and will eventually include at least some older stages, including stage 25. These grey-level, voxel models are being reconstructed from images of serial histological sections, generally from the same embryos as illustrated in *The Atlas of Mouse Development* (Kaufman, 1994). The major named anatomical components in each model embryo will be delineated in colours transparently overlaying the grey-level images. Custom-built software will allow the reconstructed embryo to be resectioned digitally in any orientation to show 'photographic' views of tissue structure and to allow each delineated named tissue to be tracked through the embryo. It will also be possible to view painted anatomical components as 3D objects using commercial software, such as AVS (Advanced Visual Systems; http://www.avs.com/), which allow a wide range of manipulations, including stereo 3D visualization using special spectacles. By delineating the named structures in the model embryos, the Atlas will thus be linked to the anatomical dictionary, making it possible to combine graphical and textual descriptions of gene-expression domains.

4.2.7.3 GXD Gene-Expression Database

This databases will store and integrate data from assays of gene-expression *in situ* and in homogenized tissue (including RNA *in situ* hybridization, immunohistochemistry, Northern and Western blots, RT-PCR, RNAase protection, cDNA arrays, etc.). The database will be available via the WWW. GXD will be fully integrated with the Mouse Genome Database (MGD) to foster a close link to genetic and phenotypic data of mouse strains and mutants. Numerous pointers to other relevant databases will place the gene-expression information into the wider biological context. Data will be acquired by annotation from the literature by database editors, and by direct transfer of expression data from laboratories. Expression patterns will be described using the controlled vocabulary of the Mouse Embryo Anatomy Database and, for *in situ* studies, digital images of original expression data. Planes of sections will be annotated graphically on standardized 2D cartoons. The electronic data-submission system will feature selection lists, on-line links to databases, and electronic validation tools to facilitate standardized data annotation and cross-referencing. The database will support complex textual queries, such as combinatorial queries, and searches on data additional to gene-expression. The images of original data will be indexed via terms from the anatomy database to provide direct links between the expression domains observed and the standardized anatomical descriptions, and to make the images accessible to global text queries. The GXD and the electronic submission system are described in more detail in Ringwald *et al.* (1997).

4.2.7.4 GXD Index

The GXD database editors currently identify all newly published research articles documenting data on endogenous gene-expression during mouse development. In a first step towards annotation of the data in the GXD, these articles are indexed with regard to authors, journal, genes and embryonic stages analysed, and expression assays used. The index is updated daily, and is available in searchable form via the WWW (Table 12.1, MGEIR, Jackson site).

4.2.7.5   Graphical Gene-Expression Database (GGED)

This database will store graphical gene-expression data using the 3D images of the Mouse Atlas as a spatiotemporal framework. Data will be submitted by individual users via a WWW, Java-based, electronic interface employing custom-designed software to map the gene-expression domains to the model embryo reconstructions from the Atlas. Submitted entries will be checked by the database team. Custom software will allow users to enter data by a combination of semiautomatic and manual graphical methods and text. Additional software will allow users to analyse their data locally, for example to identify regions of expression by name, to view expression domains in 3D relation to the structure of the embryo, and to share data with collaborators. Data entered locally will be transmitted via the WWW for submission to the public database. Similarly, graphical queries will be made locally and sent to the remote database; the results will then be returned to be analysed in the context of the Mouse Atlas. The GGED database will be fully integrated with the GXD, which will contain ancillary data for each entry in the graphical database. This integration will enable users to access a wide range of information and to combine textual and image methods to enter data and query the database.

4.2.8   Other Databases Containing Gene-Expression Data for the Mammal

Large amounts of data on expressed sequences are made publicly available by dbEST and TDB (TIGR Databases, see Table 12.1). cDNA sequences constitute expression information, and the frequency of cDNA clones in non-normalized libraries can provide limited, quantitative expression information. More importantly, however, these resources and the efforts of the IMAGE consortium (http://www-bio.llnl.gov/bbrp/image.html) and the Cancer Genome Anatomy Project (http://www.ncbi.nlm.nih.gov/ncicgap/) are extremely productive in gene discovery and provide molecular reagents to study gene expression in more detail.

TBASE stores published and unpublished data on transgenic animals and targeted mutations. The database includes phenotypic information, some of which is relevant to gene-expression. The information is given in free-text format. Entries are made by TBASE staff who request verification and completion by the authors. Searches are by text term, and complex combinatorial searches that include aspects of the methods used, dates, etc. are possible. The results of a search can be given as a table or a list.

At least two databases are planned for *in situ* gene-expression information from human embryos. One of these, based at the Department of Human Genetics, University of Newcastle-upon-Tyne, will be accessible publicly via the WWW, and will include textual and image data. It is planned, in the longer term, to make this database interoperable with the MGEIR and, to this end, the human anatomical nomenclature system (see Table 12.1) that is being built for this database at the Department of Anatomy, University of Edinburgh, will be compatible with the Mouse Embryo Anatomy Database. Gene-expression information will also be held in a database of human embryos based at the Institute of Child Health, London, and associated with a graphical atlas of human embryonic development being built at the University of St Andrew's. At present, the Institute of Child Health database is not publically available, being concerned mainly with management of the material; in the longer term, however, a public database is planned which will hold images of gene-expression patterns *in situ* as well as images and textual descriptions documenting gene-expression results from homogenized material. The textual aspects of this database will, as far as possible, be made compatible with the Mouse and Zebrafish databases through the use of the standard human embryo anatomical nomenclature being developed in association with the MGEIR. Neither of the proposed public human embryo databases is yet operational.

4.2.9   Kidney Development Database

This database stores gene-expression information relating to the developing kidney. Data are organized in tabular form. Similar databases are being constructed for other organs that develop by arborization of an epithelium, for example salivary gland, lung and mammary gland (these are included in the Organogenesis Database; Table 12.1), and a master index is available from the Organogenesis Database page which points to the relevant data tables. The Kidney Development Database and the Organogenesis Database can be accessed on the WWW. Data are entered by the database team from information sent by users via the WWW or by e-mail. The Kidney Development Database contains more than 90% of currently available information on gene-expression, mainly from mouse and *Xenopus*, but data from zebrafish and other animals are included. The database contains tables of text data and some original images of gene-expression patterns, as well as data on mutant phenotypes and the effects of pharmacological treatments on organ development in culture. Published data are flagged. The data are arranged by tissue type, but not spatially mapped. Queries are supported by a number of possible indexing strategies, and more sophisticated searches are planned. This database is operational.

4.2.10   Tooth Database

The Tooth Database stores information on gene-expression during tooth development in the mouse, rat and human. The database is available via the WWW. Data include limited information on probes and antibodies, and distinguish different types of assays. Bibliographical data is included. The data are in text form and can be searched by gene type, assay type, stage and region of expression. Searches lead the user to tables showing, for example, the genes expressed at different stages of tooth development. By clicking on a particular gene, the user is provided with a set of standard diagrams illustrating the expression in different tissues (including expression in parts of tissues). Lists of genes expressed in any particular tissue can also be obtained by clicking on the appropriate tissues indicated in diagrams of the developing tooth. This database is operational.

5   THE FUTURE OF GENE-EXPRESSION DATABASES

In the present phase of data collection these databases are likely to be used mainly to store, collate and scan information. As more of genes are identified and their expression patterns characterized and entered into the databases, it will become common to use gene-expression databases to frame testable hypotheses regarding genetic interactions in development. In this phase, there is likely to be an increasing emphasis towards relating the data to our knowledge of other measures of gene function, knowledge that will include genetic and biochemical evidence about interactions between gene products, and evidence on the distribution of cellular and morphogenetic processes within the embryo. As we have noted, this will depend crucially on establishing links between different gene-expression databases as well as on establishing appropriate interconnections with other information resources of the biomedical community.

ACKNOWLEDGEMENTS

It is a pleasure to thank our colleagues, in particular Richard Baldock at the MRC Human Genetics Unit, Jonathan Bard at the University of Edinburgh Anatomy Department, Klaus Schughart at Transgene, Strasbourg, and members of the ESF Network on Gene-expression Databases for many stimulating discussions. We also thank Silvia Martinelli, Richard Durbin, Volker Hartenstein, Wilfried Janning, Monte Westerfield, Peter Vize, Christof Niehrs, Tom Strachan and Jamie Davies for providing information on their gene-expression databases.

REFERENCES

Baldock RA, Bard JBL, Kaufman MH and Davidson D (1992) A real mouse for your computer *BioEssays*, **14**, 501–502.

Bard JBL (ed.) (1997) Gene expression databases. *Sem. Cell Devel. Biol.*, **8** (5).

Davidson D and Baldock RA (1997) A 3-D atlas and gene-expression database of mouse development: implications for a database of human development. In *Molecular Genetics of Early Human Development*, (eds Strachan T, Lindsay S, Wilson D). BIOS Scientific Publishers, Oxford.

Davidson D, Baldock RA, Bard JBL, Kaufman MH, Richardson JE, Eppig J and Ringwald M (1997) Gene-expression databases. In *In Situ Hybridization. A Practical Approach* (ed. D Wilkinson). IRL Press, Oxford.

Hartenstein V, Lee A, and Toga AW (1996) A graphic digital database of *Drosophila* embryogenesis. *Trends Genet.* **11**, 51–58.

Heisenberg M and Kaiser K (1995) The flybrain project. *Trends in Neurosci.* **18**, 481–483.

Kaufman MH (1994) *The Atlas of Mouse Development* 2nd Edition. Academic Press, London.

Lockhart D, Dong H, Byrne MC, Follettie TM, Gallo MV, Chee MS, Mittmann M, Wang C, Kobayashi M, Horton H, Brown EL, (1966) Expression monitoring by hybridization to high-density olig. nucleotide arrays. *Nature Biotechnol.*, **14**, 1675–1680.

Nguyen C, Rocha D, Granjeaud S, Baldit M, Bernard K, Naquet P, Jordan BR (1995) Differential gene expression in the murine thymus assayed by quantitative hybridization of arrayed cDNA clones. *Genomics* **29**, 207–216.

Piétu G, Alibert O, Guichard V, Lamy B, Bois F, Leroy E, Mariage-Sampson R, Houlgatte R, Soularue P, Auffray C (1996) Novel gene transcripts preferentially expressed in human muscles revealed by quantitative hybridization of a high density cDNA array. *Genome Res.* **6**, 492–503.

Ringwald M, Baldock R, Bard J, Kaufman M, Eppig JT, Richardson JE, Nadeau JH and Davidson D (1994) A database for mouse development. *Science* **265**, 2033–2034.

Ringwald M, Davis GL, Smith AG, Trepanier LE, Begley DA, Richardson JE, Eppig JT (1997) The mouse gene expression database GXD. *Sem. Cell Develop. Biol.* **8**, 489–497.

Schena M, Shalon D, Davis RW, Brown PO (1995) Quantitative monitoring of gene expression patterns with a complementary DNA microarray. *Science* **270**, 467–470.

Theiler K (1989) *The House Mouse: Atlas of Embryonic Development*. Springer-Verlag, New York.

Index

Note: page numbers in **bold** refer to main discussions; those in *italics* refer to figures and tables

A

AATAAA sequence, 235
  polyadenylation signal, 230
ABF1 binding sites, 219, *220*
ACEDB database, 159–60, 174, *176*, 285
ACEMBLY program, 191
affecteds, distribution of, 52
alleles
  frequencies, *58*, *59*
  IBD/non-IBD 82, 83
analysis, *59*, *61*
  SQL representation, *64*
analysis programs, interfacing with, 64, 65–72
ARK system, 101–2
  PiGBASE 102–4
Artiodactyls
  ARK system, 101–2
  cattle, 102
  pigs, 102–4
  sheep, 104
AUG codon, 223, *224*
automated gel electrophoresis, 151

B

baboon chromosome, 13 short arm, 136, *137*, 138–42
backcross panels, 48
bacterial artificial chromosome (BAC) 151
  libraries, 45
bacterial clone map strategy, 152–3
Baylor College of Medicine Computational Resources, 38
bioinformatics, 275
BLAST program, **24–7**, 251, *252*
  algorithm selection, 25
  database selection, 25
  interface, 25, *26*
  server, 27

BLASTC program, 251
BLASTIN program, 188
BLASTN program, 199
BLASTX program, 251, 253
BLIXEM program, 199
Booroola fecundity gene, 95
breakage frequency, 115
breakage probability, RHMAP program, *123*
buried clones, 163, 166

C

Caenorhabditis elegans
  clone map construction, 153
  database, 285
(CAG)*n* repeats, 235
canonical clone, 163
cap signal, 221, *222*
carnivores, 104–5
Cattle Genome Database, 102
CCAAT box, 221
(CCG)*n* repeats, 235
cell hybrid panels, 48
cell lines, 48
CENSOR program, 236
centromeres, 94
chicken, 106
chromosome-specific libraries, 46
chromosomes
  aberrations, 94
  banding patterns, 94
  evolution, **93–4**
  long regions, 237
clipping, quality, 186–7
cloning vector identification, 187–8
coding exon prediction, 261
coding potential, content sensors, 270
Coding Recognition Model (CRM) 239
coding region
  recognition, 261–2
  scoring, 263

coding sequences (CDS) 216
codons, 216
  initiation, *222, 223,* 224
Common Assembly Format (CAF),
      human genome sequencing, 191
comparative mapping databases, 109
COMPEL database, 231
consensus band map, 160, *161, 168–9*
CONSED, editing in, 195, *196*
ConsInspector, 223
content sensors, 270
Contig Selector, 193
Contig-Joining Editor, 194
contigs
  creation of new, 165–6
  editing, 167, *168–9*
  gaps, 201
  human genome sequencing, 182
  maps, 96
  merging, 170, *171*
  ordering in human genome sequencing,
      198–9
  refining new, 166
control sets, 217
cosmid, 151
  libraries, 46
CpG islands, 216
  libraries, 46
  prediction, 238
CPROP integrated human map, 96
CROSS_MATCH program, 187, 198
CROW program, 200
cytogenetic maps, Genome Data Base
      (GDB) 96

D
Dan Jacobson's Archive of Molecular
      Biology Software, 38
data display/management, 73
database
  implementation, 61–4
  single-species, 95
depression, 85
diabetes, 85
disease
  gene size, 85
  locus, **78–81**
  recurrence risk, 80
DNA
  fragment overlaps, 118
  markers, 76
  sequence, 90

chromosome, 93
  single-chromosome libraries, 93
  tandemly repeated, 235
DNA Mapping Panel Data Sets, 98
cDNA libraries, 47
  EST generation, 206, 207
cDNA splice site prediction, 228–9
Dog Genome Project, 105
DogMap, 104–5
DOTTER program, 199, 200
Drosophila database, 281, 284
  EMBRYO 286
dynamic programming, **262–6**
  algorithm, 264–5, 269
  finite state automaton, 267
  state diagram, 267, *268*

E
EMBL database, 231, *232*
Encyclopedia of the Mouse Genome, 100
entity modelling, 56, *57*
  analysis, *61*
  genotypes, 59, *60*
  phenotypes, 59, *60*
Entity Relationship Attribute (ERA)
      model
  analysis, 49, *55*
  mapping, 6, 631
  normalization, 61
Entrez search engine, **4–12,** *13*
  accessing, 5
  divisions, 5–6
  genomes division interface, 12
  results presentation, *9*
  structural division, 12, *13*
  user interface, 11
EPD database, 231, *232*
ERPA 88
Escherichia coli
  genome, 188
  transposons, 198
ESPA (Extended Sib-Pair Analysis)
      program, 82–3
EST databases
  consensus sequence, 210
  value added, 209
  see also expressed sequence tags (ESTs)
European Collaborative Interspecific
      Backcross (EUCIB) 100–1
European Molecular Biology Laboratory
      (EMBL) 15
Eutheria, 89–90

evolution
  chromosomes, 93–4
  vertebrate, 90
exon, 216
  identifiers, 1
  prediction, 29–30, *31*
  protein coding gene structure
    prediction, 237
  recognition, 240–1
  score calculation, 263, 265
expressed sequence tags (ESTs) 89,
    205–13
  analysed data, 209–11
  analysis tools, 211–12
  BLAST database, 210
  chimeric cDNA clones, 209
  contamination, 208
  data generation, 206–7
  definition, 205
  errors, 208
  gene expression data, 290
  gene mapping, 238
  generating, 206–7
  human sequences, *see* human EST
    sequences
  ICAtools package, 211–12
  IMAGE program, 212
  mapping, 207
  Merck Gene Index project, 210
  NCBI database, 207
  radiation hybrid mapping programs,
    145
  raw data, 209
  SaniGene, 211
  sequencing, 207–9
  STACK database, 211
  TIGR assembler, 212
  TIGR EGAD 211
  TIGR HGI 211
  UniEST project, 210
  UniGene project, 210
  universal primers, 206–7
  see also human EST sequences

F
FAKII program, 190, 191
families, 56
  partially informative, 82–3
  types, 52
FASTA program, 251, 252
FGENEH program, *239*, 240–1, 263,
  271

FINISH program, 197
fish, teleost, 106–7
fluorescence, *in situ* hybridization (FISH)
  93
  EST mapping, 207
FlyBase, 286
Flybrain database, 286
Flyview database, 280, 282, 285–6
FPC **160–72**
  buried clones, 163
  canonical clone, 163
  consensus band map, *168–9*
  contig
    display, *162*
    editing, 167, *168–9*
    merging, 170, *171*
    new, 165–6
  customization, 172
  framework map, 174
    markers, 171
  installation, 163
  marker adding, 171
  new clone adding to databse, 170
  probability of coincidence score, 163
  project, 164
  sequence ready clone selection, 172
  sequence ready contig construction,
    160–7, *168–9*, 170–2
  tile path selection, 172
  V2.8 162–7, *168–9*, 170–2
frame-aware algorithm, 267, *268*
frameshift mutations, 237–8
Fugu Genome Scanning, 106–7
FunSiteP program, 233

G
GAP4 program, 189, 190, 191, 199, 200
  editing in, 193–5
  finishing in, 196–7
GC box, 221
GC splice site, 230
GC-rich isochores, 216–17
gel image processing, 182–5
GELMINDER program, 184
GenBank, 4–12
  accessing, 5
gene
  conversion, 90
  copies in mammals, 90–1
  deletions, 90
  duplication, 90
  homologous, 92, *93*

gene (*continued*)
  identification, 93
  orthologous, 92
  paralogous, 92
  promoter prediction, 261
  protein coding gene structure
    prediction, 237
  recognition tools, 238
  sequence similarity, 91–2
  signal detection, 261
  within-species variation, 91
gene conservation, **89–92**
  diverse phyla, 89
  vertebrate, 89–90
gene expression data survey, **278–84**
  annotation, 283
  assay method, 279
  data acquisition, 282
  data submission, 282–3
  developmental data, 279–80
  experimental evidence, 282
  gene name, 278–9
  linked entries within database, 283
  links to other databases, 283–4
  marker gene expression onset, 280
  method, 279
  nomenclature systems, 280
  probe specification, 279
  quantitative data, 281–2
  spatial aspects, 280–1
  temporal aspects, 279–80
  textual records, 281
  tissues, 279
gene expression databases, 275–7,
  284–92
  access, 284–5
  Java applets, 284–5
  mammals, 290–2
  phenotypic data, 277
  query capabilities, 278
  scale, 277
  scope, 277–8
  search capabilities, 278
  sequence data, 277
  splice forms, 278
  types, 277
  users, 276
Gene Family Database, 108
Gene Finder program, 29–30
gene finders, *see* exon prediction
gene finding, performance comparison,
  271–2

gene identity, estimation between species,
  90–2
Gene Map of the Human Genome,
  13–15
gene mapping, comparative, 94–5
gene order
  conservation in mammals, 94
  determination, 92–3
  organelle genomes, 94
gene structure
  chromatin structure, 215
  non-random nucleosome positioning,
    215
  nucleotide string, 215
  optimal, 269
  prediction combining sensor
    predictions, 262
genealogy, *see* pedigree
Genefinder program, 262
GENEHUNTER program, 85
GeneID program, *239*, 243–4, *245*, 271
GeneMark program, *239*, 244, 246, 247
GeneModeler program, 262
GeneParser program, *239*, 244, 246,
  263, 265, 271
Generalized Hidden Markov Model
  (GHMM) 248, 270
  see also hidden Markov Model
  (HMM)
Genethon genetic map, 14
genetic identity-by-descent
  recombination fraction, 77–8
  sib pairs, 75–7
  see also identity-by-descent (IBD)
genetic linkage mapping, 92
GeneWise program, 263
Genie program, *239*, 248, 263, 270, 272
GenLang program, *239*, 248, 252
Genome Database (GDB) 20–1, 22, 95
  display, *22, 23*
  Gene Family Database information
    presentation, 108
  gene nomenclature, 96
  home page, 20, *21*
  map recovery, 21
  searches, 20–1
genome databases, **3–24**, 39
  GDB 20–1, 22
  Gene Map of the Human Genome,
    13–15
  species-specific, 22–4
  SRS 15–20

WWW Virtual Library, 3–4
GenomeInspector program, 231
genomic libraries, 42–3, 45–6
    PCR pools, 42–3
genomic resource availability, 41, *42, 43*
Genomic Segment, 95
genotype, 54, 59
    data management, 49–50
    entity modelling, *60*
    SQL representation, *63*
GENSCAN program, 270, 272
GenView program, 241
GETLANES program, 184
GRACE program, 185
GRAIL (Gene Recognition and
    Analysis/Assembly Internet Link)
    30, *31*, 238–40, 263
    human HUMNTRIII sequence, *240*
GREAT program, *239*, 249
GXD Gene Expression Database, 288,
    289
GXD Index, 100

H
haemoglobins, aberrant human, 91
Hamming-Clustering network technique,
    222, 230, 235
haplotypes, *51*, 54
hidden Markov model (HMM) 267–9
high-throughput DNA sequencing, 151
    automated gel electrophoresis, 151
    shotgun sequencing, 151, *152*
*Hin*dIII sites, 155
HMMER program, 198
HMMgene program, 269, 270, 272
Homeobox Page, 108
homologous vertebrate genes
    (HOVERGEN) 107–8
homology, 92, *93*
    detection, 237
    search, 253
Hox genes, 90
HSPCRUNCH program, 251
HSPL program, 228
human embryos
    databases, 290, 291
    in situ gene expression, 290
human EST sequences, 205–6
    analysis tools, 211–12
    biological resources, 212
    cDNA libraries, 206, 271
    publicly available data, 209–11

see also expressed sequence tags (ESTs)
human gene structure prediction,
    215–54
    coding region methods, 236–7
    CpG island prediction, 238
    donor splice site, 224–7
    ESTs, 238
    exon recognition, 240–1
    FGENEH *239*, 240–1
    frameshift mutations, 237–8
    functional sites in nucleotide
        sequences, 219, *220, 221*–30
    gene recognition tools, 238
    GeneId, *239*, 243–4, *245*
    GeneMark, *239*, 244, 246, *247*
    GeneParser, *239*, 244, 246
    Genie, *239*, 248
    GenLang, *239*, 248
    GenView program, 241
    GRAIL 238–40
    GREAT *239*, 249
    Hamming-Clustering network
        technique, 235
    homologous protein sequence, 253
    homology
        detection, 237
        search, 253
    initiation codon, 222
    introns, 234
        splicing, 224
    MZEF *239*
    ORFGene program, 242
    polyadenylation signal, 230
    potential protein coding analysis,
        251–2
    problems, 236–8
    PROCRUSTES *239*, 246–8
    protein coding
        gene structure, 236–51
        regions, 233–4
    repeated elements, 253
    repeated regions, 235–6
    RNA polymerase II promoter elements,
        221
    RNA-coding, 252
    sequence analysis, 252–4
    SORFIND *239*, 249, *250*
    splice site
        analysis/recognition, 224
        methods, 237
    transcription factor binding sites, 219,
        220, 221–30, 253

human gene structure prediction
      (*continued*)
   translation initiation sites, 223–4
   3' UTR 235
   WebGene, *239*, 241–2, *243*
   weight matrices, 221
   Xpound, *239*, 251
human genome, gene density, 205
Human Genome Mapping Project, 275
human genome sequencing, **181–203**
   ACEMBLY program, 191
   assembling readings, 195
   assembly
      interchange formats, 191–2
      pragmatics, 190–1
   base representation, 192
   BASS program, 185
   BLASTIN program, 188
   BLASTN program, 199
   BLIXEM program, 199
   cloning vector identification, 187–8
   Common Assembly Format (CAF) 191
   CONSED program, 195, *196*, 199
   contaminant removal, 188
   contig, 182
      construction, 189
      gaps, 195, 201
      ordering, 198–9
   Contig Selector, 193
   Contig-Joining Editor, 194
   CROSS MATCH program, 198
   CROW program, 200
   data transfer, 185
   DOTTER program, 199, 200
   editing, 193–5, *196*
      in CONSED 195, *196*
      GAP4 program, 193–5
   error rates, 192
   experiment files, 188–9
   FAKII program, 190, 191
   feature identification, 197–9
   FINISH program, 197
   finished sequence, 200–1
   finishing, 182, 192–3, 198
      automated, 196–7
   GAP4 program, 189, 190, 191,
      193–5, 199, 200
      finishing in, 196–7
   gel image processing, 182–5
      backround subtraction, 183
      base calling, 184
      colour deconvolution, 183
      lane tracking, 183
      mobility correction, 184
      peak normalization, 184
      quality estimation, 184
      trace extraction, 183
      trace smoothing, 184
      Unix-based gel image processing,
         184–5
   GELMINDER 184
   GETLANES 184
   GRACE 185
   HMMER program, 198
   informatics, 181
   misassembly, 197
   PCOP program, 200
   PHRAP program, 185, 186, 190, 191,
      192, 195
   PHRED program, 184, 187, 195
   polymorphisms, 201
   PREGAP program, 188
   PRINTREPEATS program, 199
   quality control, 200
   RANCL program, 182
   repeat sequences, 197–8
   REPEATMASKER program, 198
   sequence assembly, 182, 189–92
   sequence preprocessing, 185–9
      quality clipping, 186–7
   sequencing vector identification, 187
   shotgun strategy, 182
   software, 181–2
   subclones, 182
   tagging, 199
   tandem repeats, 201
   TED program, 186, 187
   TREV program, 186
   Unix-based preprocessing packages,
      188
   viewing features, 199
human genome structure, **216–18**
   chromatin fibre synonymous
      substitution, 217
   coding measure evaluation, 218
   codons, 216
   control sets, 217
   CpG islands, 216
   exon, 216
   GC-rich isochores, 216–17
   prediction accuracy evaluation,
      217–18
   primary structure, 216
   protein density, 216

RNA coding sequences, 216
human HUMNTRIII sequence, *240, 241, 243, 245, 246, 247, 250*
human pedigree analysis, 49
human Xq chromosome, 105
human-mouse homologies, 109
hypoxanthine phosphoribosyltransferase (HPRT-) deficiency, 113

I
ICAtools package, 211–12
identity-by-descent (IBD) **75–7**
  alleles, 82, 83
  conditional distribution for sib-pair, 77, *78*
  counting methods, 81–2
  distribution for affected sib-pair, 78–81
  partially informative families, 82
  probabilities, 79
  statistics, 81–2
identity-by-state (IBS) 75–7
IMAGE (Integrated Molecular Analysis of Gene Expression) consortium library, 47
IMAGE program, 212
Integrated Genomic Database (IGD) 49
integrated maps, 96
International Society for Animal Genetics (ISAG) 105
internet connection, 2–3
INTRON program, 229
introns, 234
  splicing, 224
Ion Channel Database, 284
irradiation and fusion gene transfer, 114
IUPAC-IUB standard codes, 221

J
Java applets, 1, 21, *23*
  gene expression databases, 284
  promoter regions, 231

K
Kidney Development Database, 282, 283, 291

L
libraries
  bacterial artificial chromosome, 45
  chromosome-specific, 46

cosmid, 46
CpG island, 46
cDNA 47
IMAGE 47
P1 artificial chromosome, 45–6
phage, 46
YAC 45
likelihood methods of sib-pair analysis, 83–4
LINKAGE 64, *65–9*
  MAPMAKER/SIBS program, 84
linkage analysis, 54–5
  disease gene size, 85
  using affected sib-pairs, 75
linkage group
  conservation, 92–4
  cytogenetic anchoring to chromosomes, 104
linkage maps, Genome Data Base (GDB) 96
Location Database (LDB), integrated human map, 96
lod score, 116
  method, 85
  pairwise, 116
  RHMAP program, *123*
long interspersed elements (LINEs) 235

M
malignant hyperthermia, *95*
mammals
  evolution, 90
  gene copies, 90–1
  gene order conservation, 94
MAPMAKER/SIBS program, 84, 88
mapping programs, 1
Mapview program, 96
marker, *53–4*
  models, *53–4, 58–9, 63*
  set, *61*
  SQL representation, *63*
  systems, *58–9*
Markov chain, inhomogenous, 263, 270
marsupials, 105
MatInspector, 223
MATRIX SEARCH database, 223
maximum likelihood estimation method, 269–70
MBx database, 101
MEF2 database, 219
Mendel database, 108–9

Mendel's law of segregation, 76
Merck Gene Index project, 210
Metatheria, 89, 90
MGEIR database, 184, 280, 281, 282, 283, **288–90**
    composition, 288
    Graphical Gene Expression Database (GGED) 288, 290
    GXD Gene Expression Database, 288, 289
    GXD Index, 289–90
    Mouse Atlas, 288–9
    Mouse Embryo Anatomy Database, 285, 288
    Zebrafish database link, 287
microsatellites, 53–4
minimum breaks (RHMINBRK) method, 119
misassembly, human genome sequencing, 197
Mn/Fe superoxide dismutases, 90
    partial sequences, *91*
model-based analysis, 50, 52–4
monotremes, 105
Mouse Atlas, 288–9
mouse chromosome 2 143
    RADMAP *144*, 145
Mouse database, 291
Mouse Embryo Anatomy Database, 285, 288
Mouse Gene Expression Information Resource, *see* MGEIR database
Mouse genome, EUCIB high-resolution genetic mapping, 100
Mouse Genome Database (MGD) 96, **97–8**, *99*, **100**
    accessing, 98
    contents, 97–8
    DNA Mapping Panel Data Sets, 98
    grid display, *99*
    GXD Index, 100
    homology information, 98
    links, 97
    map display, 100
    marker records, 98
    phenotype records, 97, 100
    query forms, 98, 100
Mouse Locus Catalog (MLC) 100
MPSRCH program, 252
MultiMap, 144–5
myogenin site database, 219
MZEF program, *239*

N
NCBI
    gene map, *14*
    UniGene set, 15
NCBI database
    expressed sequence tags (ESTs) 207
    model organism cross-referencing, 89
NCBI home page, 4–5, 13
    Research Tools Page, 14–15
NCSA Biology Workbench, 25, **30–4**
    analysis performance, 33, *34*
    analysis selection, 33
    analytic subdivisions, 31–2
    importing sequences, 33
    sequence alignment, *35*
    sequence selection, 33
NETGENE program, 227, 228
neural networks, 227, 263, 266
    Genie program, 248
nuclear three-generational families, 52
nucleotide
    database, 6, 7
    repeats, 53
nucleotide sequences
    functional regions, **230–1**, *232*, **233–6**
    functional sites, **219**, *220*, **221–30**
    promoter regions, 231, *232*, 233

O
obligate breaks, 116, 117, *118*
Online Cytogenetics of Animals (OCOA) 94
Online Mendelian Inheritance
    in Animals (OMIA) 107
    in Man (OMIM) 95, 96
Open Data Base Connectivity (ODBC) 73
open reading frame, sliceable (SORF) 249
ORFGene program, 242
organelle genomes, gene order, 94
Organogenesis Database, 291
orthologies, 92

P
P1 artificial chromosome (PAC) 151
    libraries, 45–6
paralogies, 92
parents, 56
parsimony, 94
PCOP program, 200
pedigree, 50, *51*, 52
    data display, 73
    data management, 49–50, 73

entities, 56, 57
set, *60*
SQL representation, *62*
structure, *55–8*
Pedigree/DRAW *70–2*
Pedro's BioMolecular Research Tools, 38
Perl commands, 143
phage libraries, 46
phenotypes, 54, 60
  entity modelling, *60*
  SQL representation, *63*
PHRAP program, 185, 186, 190, 191,
    192, 195
PHRED program, 184, 187, 195
physical mapping tools, 34–8
PiGBASE 102–4
poly(A) site prediction, 230, 235
polyadenylation signal, 230
  Hamming-Clustering network
    technique, 230
polymerase chain reaction (PCR) 53
polymorphic markers, 92
polymorphisms, 53, 54
  human genome sequencing, 201
polypyrimadine tract, 227
PREGAP program, 188
primates, single-species database, 95–7
primer banks, 47
primer picking, 27–9
  PCR primer pair, 27, 29
PRIMER program, 27
probe banks, 47
PROCRUSTES program, *239*, 246–8, 263
promoter regions, 231, *232*, 233
  COMPEL database, 231
  EMBL database, 231, *232*
  EPD database, 231, *232*
  FunSiteP program, 233
  GenomeInspector program, 231
  Java applet, 231
  PROMOTER SCAN program, 233
  SWISSPROT database, 231
  TRANSFAC database, 231, *232*
  TRRD database, 231
  WEBGENDB database, 231
PROSITE database, 252
PROSRCH program, 252
protein
  analysis of potential coded by
    predicted genes, 251–2
  coding regions, 233–4
  databases, 251–2

families
  homologous vertebrate genes
    (HOVERGEN) 107–8
  Mendel, 108–9
homologous sequence, 253
structure motif finders, 1
protein coding gene structure prediction,
    236–51
  exons, 237
  FGENEH 240–1
  gene structure prediction, 236–8
  GeneID 243–4, *245*
  GeneMark, 244, 246, *247*
  GeneParser, 244, *246*
  genes, 237
  Genie, 248
  GenLang, 248
  GRAIL 238–40
  GREAT 249
  MZEF 251
  PROCRUSTES 246–8
  SORFIND 249, *250*
  WebGene, 241–2
  Xpound, 251
protein-protein interactions,
    transcription regulation, 231
pseudo-buried clones, 163, 166
PSITE program, 252

Q
quantitative trait loci (QTLs) 95

R
radiation hybrid mapping, 92–3,
    113–15
  breakage frequency, 115
  comprehensive maps, 116
  concepts, 115–18
  definitions, 115–16
  DNA fragment overlaps, 118
  EST mapping, 207
  framework maps, 116
  human genome, 115
  obligate breakpoints, 116, 117, *118*
  panel size, 114
  programs, 114
    ESTs, 145
    RADMAP 144–5, 147
    RHMAP 114, 119, *120*, 121–32,
      133, 145, 147
    RHMAPPER 134–6, *137*, 138–44,
      145, 147

radiation hybrid mapping (*continued*)
  programs (*continued*)
    whole genome, 145
  radiation dosage, 117–18
  radiation hybrid ploidy, 118
  retention, 117
  RHMAP 119, *120*, 121–32, *133*
  selectable marker, 113
  whole genome, 114
  Z-RHMAPPER 157, 159
radiation hybrids, 48
  maps, 96
RADMAP 144–5, 147
  analysis output, *146*
  analysis package, 114
  mouse chromosome, 2 *144*, 145
RANCL program, 182
RATMAP database, 101
recombination fraction, genetic identity-
      by-descent, 77–8
relational model mapping, 61, 63
repeat sequences, human genome
      sequencing, 197–8
repeated elements, 253
repeated regions, 235–6
  CENSOR program, 236
  SINE element prediction, 236
REPEATMASKER program, 198
Research Tools Page, 14–15
retention
  frequency, 116
  models, 117
  rate, 117
RH2PT 125–6, 127
RHMAP program, 114, 115, 119,
      121–32, *133*, 145
  actual retention rate, 132
  analysis output, *146*
  breakage probability, *123*
  data, 119, *120*, 121
  distance estimates, 132
  influential hybrids, 129
  linkage statistics, *123*
  locus retention probability, *122*
  lod scores, *123*
  maximum likelihood (RHMAXLIK)
      methods, 119, 129–32, *133*
  minimum breaks (RHMINBRK)
      method, 119, 124–9, 130, 132
  orders, 128, 129
  retention
    models, 119, 147

patterns for markers, *120*
probability, 132
rate for markers, *128*
two-point analysis (RH2PT) program,
    119, 121–4, 125–6, 127
two-point linkage groups, *124*
RHMAPPER program, 114, 115, 134–6,
    137, 138–44, 145, 147
  analysis output, *146*
  baboon panel, 136, *137*, 138–42
  customizing, 143
  data, 135–6
  directory, 135–6
  framework map, 134, 138
  lod scores, 138, 139
  map printing, 141–2
  mouse chromosome 2 *142*, 143
  parameters, 136
  Perl commands, 143
  placement map, *140*, 141, 142
  via the Web, 144
RHMAXLIK 129, 130, 132
RHMINBRK program, 124–5
ribosomal RNA gene prediction, 252
ribosome skipping model, 223
RNA polymerase II promoter elements,
    221
RNA-coding gene structure prediction,
    252
tRNA genes, 236
  prediction, 252
RNASPL program, 229
rodents
  mouse genome, 97–101
  RATMAP 101
  single-species database, 97–8, 99, 100–1
ryanodine receptor gene (*RYR1*) 95

S
Saccharomyces cerevisiae 188
  clone map construction, 153
SAM program, 101, 156–7, *158*
SaniGene, 211
SBASE library, 251
sequence annotators, *see* exon prediction
sequence assembly, 182, 189–92
sequence ready clone maps, **151–5**
  ACEBD data base, 174, *176*
  ACEDB database, 159
  bacterial clone contigs, 155
  bacterial clone map strategy, 152–3
  contig construction, 154

contig merging, 174
coordination, 172–4, *175–6*
data transfer, 173
framework marker file, 174, *175*
implementation, 173–4, *175–6*
long-range, 155, 156–7, *158*, 159
process tracking, 172–4, *175–6*
properties, 153
SAM program, 156–7, *158*
software, 155–6
strategy, 153–5
STS data, 174
STS markers, 154, 155
STS-content database, 159–60
tiling paths, *152*, 153, 155
YAC/STS maps, 156–7
Z-RHMAPPER 157, 159
sequence ready clones, selection, 172
sequence ready contig construction using FPC **160–7**, *168–9*, **170–2**
consensus bands map, 160, *161*
FPC V2.8 162–7, *168–9*, **170–2**
fragment set generation, 161
shared fragments between clones, 160
sequence ready contigs, 174
sequence ready tile path selection, 172
sequence similarity search engines, 1
SheepBase, 104
shotgun sequencing, 151, *152*, 182
sib-pairs, 52
analysis of affected
alleles, 82, 83
computer programs, 87–8
likelihood methods, 83–4
limitations, 84–5
methods, 81–4
other affected relative pairs, 85
partially informative families, 82–3
strengths, 84–5
unaffected family members, 85
identity-by-descent (IBD) 75–7
distribution, 78–81
recombination fraction, 77–8
identity-by-state (IBS) 75, 77
linkage analysis, 75
SIGNAL SCAN database, 223
signal sensors, 263, 270
SINE element prediction, 236
single-species databases
artiodactyls, 101–4
carnivores, 104–5
chicken, 106

marsupials, 105
monotremes, 105
rodents, 97–101
teleost fish, 106–7
somatic cell hybrid panels, 48
SORFIND *239*, 249, *250*
species-specific databases, 22–4
splice site
acceptor, 227–8
prediction errors, *228*
branch point signal, 228, *229*
donor, 224–7
consensus sequence, *225*, 226
neural network approach, 227
prediction errors, *226*
sequence patterns, 225
false, 224
GC 230
INTRON program, 229
non-canonical, 230
prediction, 261
in cDNA 228–9
problems, 229–30
program, 225–6
recognition, 261–2
RNASPL program, 229
sensors, 268
signal sensors, 270
weight matrix approach, 224, *226*
spliceosomes, 230
SpliceView, 226
SPLINK program, 84, 88
SQL 61, *62*, 63, *64*
SRS (Sequence Retrieval System) **15–20**
accessing, 15–16
BLAST search interface, 25
databases linked, *16*
home page, 16
links, 18–19
search results options, 19–20
text fields, 16–17
STACK database, 211
Stanford RH map, STS placing, 37–8
state models, 266–70
Structured Query Language (SQL) 50
STS (sequence tagged site) 27–8
amplification, 35
mapping of new, 34–5
mapping results, 36
markers, 154, 155
YAC contigs, 36–7
sushi search, 6, 7, 8, 10–12

SVEC_CLIP program, 187
SWISSPROT database, 231

T
tandem repeats, 201
TATA-box, 221, 222
TBASE database, 280
telomeres, 93–4
tentative human consensus (THC)
    sequences, 211
termination site recognition, 262
The Comparative Animal Genome
    Database (TCAGDB) 109
The Institute for Genomic Research
    (TIGR)
    assembler, 212
    Expressed Gene Anatomy Database
        (EGAD) 211
    Human Gene Index (HGI) 211
thymidine kinase deficiency (TK-) 113
Tilapia Genome Project, 106
Tooth Database, 281, 283, 292
trait models, 57, 58
    SQL representation, 62
traits, 52–3, 58
    entity modelling, 57
    modelling expression, 53
    qualitative models, 53
    SQL representation, 62
transcription factor binding sites, 219,
    220, 221–3
    classes, 222–3
    databases, 223
transcription regulation, protein-protein
    interactions, 231
Transcriptional Factor Database (TFD)
    219
TRANSFAC database, 219, 220, 221,
    231, 232
translation initiation, 262
    sites, 223–4
TREV program, 186
TRRD database, 231

U
UniEST project, 210
unigene databases, 13–15
UniGene project, 210
    dataset, 209
universal primers, 206–7
Unix-based gel image processing, 184–5
Unix-based preprocessing packages, 188

URL (Universal Resource Locator) 1,
    39–40
US Pig Gene Mapping website, 104
5 UTR 216
3 UTR 216, 235

V
vector clipping/identification, 187
VECTOR_CLIP program, 187, 188
VEIL program, 268, 269
vertebrate comparative mapping
    database, 109
Viterbi algorithm, 269

W
Web browsers, 2, 39
WEBGENDB database, 231
WebGene, 239, 241–2, 243
Whitehead Institute Biocomputing Links,
    38
Whitehead radiation mapping service,
    36–7
World Wide Web, **1–2**
    analytic tools, 24–38, 40
    database data submission, 282
    equipment, 2–3
    EST resources, 213
    gene expression databases, 284
    resource availability, 41, 42, 43
    RHMAPPER use, 144
    software finding, 202–3
    suppliers, 41, 44
    Virtual Library, 3–4

X
Xenopus Molecular Marker Resource,
    280, 284, 287
Xp chromosome, 105
Xpound, 239, 251
Xq chromosome, 105

Y
YAC/STS maps, 156–7
yeast artificial chromosome (YAC)
    contigs, 36–7, 113
    libraries, 45

Z
Z-RHMAPPER 157, 159
Zebrafish database, 282, 284, 287,
    291
Zebrafish Genome Project, 106